I0045900

The Enigma of the Crookes Rad
by Stefan Hollos and J. Richard Hollos

Abrazol Publishing
an imprint of Exstrom Laboratories LLC
662 Nelson Park Drive, Longmont, CO 80503-7674 U.S.A.

Publisher's Cataloging in Publication Data
Hollos, Stefan
The Enigma of the Crookes Radiometer / by Stefan Hollos and J. Richard Hollos
p. cm.
Includes bibliographical references
Paper ISBN: 978-1-887187-44-2
Library of Congress Control Number: 2022936904
1. Physics – History 2. Crookes, William, 1832-1919
3. Radiometers 4. Kinetic theory of gases
5. Vacuum technology
I. Title. II. Hollos, Stefan.
QC7.H65 2022
530.09 HOL

About the Cover:

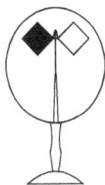

Contents

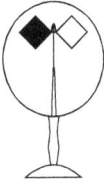

Pictured above is a Crookes radiometer of the type you can buy in museum gift shops and toy stores all over the world. This particular one was purchased in the gift shop of the Museum of Science and Industry in Chicago, sometime in the late 80's. It still works as well as the day it was purchased, over 30 years ago. All you need to do is shine light on it and the vanes will start to turn.

1

It is a fairly simple device. Four vanes mounted symmetrically around a tube shaped spindle that sits upside down on a needle point, so that it can turn with very little friction. One side of each vane is black while the other side is white or silvered. All of this is enclosed in a clear glass bulb from which most, but not all, of the air has been removed.

There have been many variations on this simple design. Some involve variations on the number and shape of the vanes. Others involve enclosures of varying size and shape. In most cases we have something that resembles a tiny windmill or paddle wheel in an enclosure.

When you shine light on the vanes they start to turn, so an obvious conclusion is that the light is somehow exerting pressure on the vanes, causing them to turn. That was also the initial conclusion of Sir William Crookes (1832-1919), a British chemist and physicist, who first created one of these devices. It is such a common explanation for how the device works that it is often called a light mill.

The problem is the explanation is wrong. The light only indirectly provides the energy needed to make the vanes turn. It does so by heating up the black side of the vanes more than the white. This temperature difference is what powers the movement of the remaining air molecules in the bulb, in such a way as to make the vanes turn. In fact, if you could take all the air out of the bulb it would stop turning.

In spite of the apparent simplicity of the device, it was surprisingly hard for physicists to figure out exactly how

Figure 1: Caricature of Sir William Crookes, circa 1902, published in Vanity Fair, May 21 1903. Public domain image courtesy of Wikimedia Commons at https://commons.wikimedia.org/wiki/ File:Sir_William_Crookes_1902.jpg

it worked. For awhile it was indeed an enigma. The first public demonstration, by William Crookes, of the effect on which the radiometer is based, was at a 1874 meeting of the Royal Society. It attracted the interest of some of the best scientists of the time. The great Scottish physicist James Clerk Maxwell (1831-1879) and the Irish-born fluid dynamicist Osborne Reynolds (1842-1912) both spent time trying to explain how it worked.

It was not until 1879 that they came up with what is now generally considered to be the correct theoretical explanation. Both theoretical and experimental work on the radiometer continued off and on for another 50 years. Even German-born physicist Albert Einstein (1879-1955) published a paper on it in 1924 [1].

A renewed interest in the physics behind the radiometer began in the 1980s and continues to this day. This is mostly driven by work on things like microelectromechanical systems (MEMS) and atomic force microscopy (AFM), where the same kind of physics is involved. It is also related to the movement of particulate matter in the upper atmosphere. Some people have even proposed exploring the upper reaches of the atmosphere using vehicles powered by the same forces that make the radiometer vanes turn [2].

Much research has been done on the radiometer. Crookes

---

[1] "Zur Theorie der Radiometerkräfte" https://einsteinpapers.press.princeton.edu/vol14-doc/549 to 558, translated as "On the Theory of Radiometer Forces" https://einsteinpapers.press.princeton.edu/vol14-trans/318 to 322

[2] https://bargatin.seas.upenn.edu/

Figure 2: Albert Einstein in 1921, who 3 years later published a paper on the radiometer. Public domain image courtesy of Wikimedia Commons at
https://commons.wikimedia.org/wiki/
File:Einstein_1921_by_F_Schmutzer_-_restoration.jpg

Figure 3: James Clerk Maxwell worked on understanding the radiometer just before he died in 1879. Public domain image courtesy of Wikimedia Commons at https://commons.wikimedia.org/wiki/ File:James_Clerk_Maxwell.png

Figure 4: Osborne Reynolds (from 1904 painting) worked on understanding the radiometer in the 1870's. Public domain image courtesy of Wikimedia Commons at https://commons.wikimedia.org/wiki/ File:OsborneReynolds.jpg

himself made a countless number of different radiometers. He varied every conceivable characteristic of the device, to determine its effect on performance. The experimental and theoretical work done by Reynolds and Maxwell in the 1870's did convince most physicists that the basic physics behind the device was understood, and interest in it started to wane by the end of the century. This was really just a case where possible avenues of further investigation ran into a dead end, due to limitations in experimental equipment available at the time.

The German physicists Walther Gerlach (1889 – 1979) and Wilhelm Westphal (1882 – 1978) began investigating the radiometer in the 1920's to see if it could indeed be used to measure radiation pressure, given a low enough vacuum. In a paper Gerlach published in 1923[3] he opens with the statement:

"As is well known, there is no complete theory of the radiometer available."

This is a somewhat surprising statement given that it comes almost 50 years after Crookes first began his investigations. But the fact is that, even today, it is hard to explain exactly how the device works in simple concise terms. While the basic physical principles behind its operation are not really up for debate, exactly how they apply to a particular device is not always clear. This has led a lot of current researchers to use computer simulations, to better understand what is going on.

---

[3]Untersuchungen an Radiometern. W. Gerlach, H. Albach, Zeitschrift für Physik 14 (1923), 285-290.

Figure 5: Wilhelm Westphal in 1935, who wrote a series of papers on the radiometer in 1920-1921.
Image from Wikimedia Commons at
https://commons.wikimedia.org/wiki/
File:Westphal,Wilhelm_1935_Stuttgart.jpg
under the Creative Commons Attribution 3.0 Unported (CC BY 3.0) license at
https://creativecommons.org/licenses/by/3.0/deed.en
courtesy of Gerhard Hund, from his father Friedrich Hund.

What we know for sure is that it is not light pressure that makes the vanes turn. Light does exert pressure on the vanes but it is nowhere near enough to make them turn. The effect of the light is to heat up the vanes and the surrounding glass bulb. Visible light passes through the glass bulb and heats up the black side of the vanes more than the white. Infrared radiation[4] on the other hand is absorbed by the glass bulb, heating it up.

When the residual gas molecules in the bulb interact with the vanes and the inner surface of the bulb, their kinetic energy will change. On average they will rebound from a hotter surface with more kinetic energy than from a colder surface. Naively, one would then expect the hotter black surface to experience a greater pressure from the rebounding molecules than the white surface and that this pressure difference is what turns the vanes. To a certain extent this is true but the details are a bit more complicated. There are also other processes at work that we will discuss later in the book.

To exactly explain what is going on requires the use of the kinetic theory of gases. For a particular radiometer design, computers are often used to simulate the gas dynamics and explain exactly how it makes the vanes turn. Such simulations are beyond the scope of this book. We will restrict ourselves to a more general, semi-quantitative description of how the radiometer does and does not work.

The outline of the book is as follows. We begin with

---

[4]Infrared radiation has wavelengths too long for the human eye to see, but it can be sensed as heat by the skin. All warm objects emit infrared radiation.

a history of the origin of the device and the bewildering number of experiments that Crookes performed. He created radiometers of many different kinds, shapes and sizes. The common radiometer is but one of many possible ways to design a radiometer. This is followed by a description of how it works in general terms, that we think can be understood by people with only a very modest understanding of physics. Some of the more technical details have been worked out in a problems section which requires a bit more of a physics background.

The main protagonist in this story is of course Sir William Crookes who is an interesting individual in his own right. For those interested in learning more about him there is a short biography and timeline of his life in the appendix. A bibliography with references where you can learn more about the radiometer is also included.

Herriman, Hollos and Hollos

The origin of the radiometer begins with Crookes' discovery of thallium[5] in 1861, using the then new technique of spectral analysis. This discovery got him a lot of attention and he spent the next decade trying to precisely determine its atomic weight. Since thallium is a heavy element and is easily oxidized, a precise determination of its atomic weight at the time was difficult. To get better accuracy he decided to weigh his thallium samples in a vacuum and in the process discovered a strange new phenomenon.

When a sample was hot it appeared to weigh less than when it was cold. At first he thought he may have discovered a new relationship between heat and gravity and this launched him into a long series of experiments to more accurately measure the effect. Crookes was a very thorough and energetic experimenter. In 1874 he published the first of several papers describing his numerous experiments. The paper was titled "On Attraction and Repulsion Resulting from Radiation," [6] and it describes the devices he built to help him isolate and study the effect he was seeing in the weight measurements.

The first device that gave him good repeatable results is shown in figure 6. It is basically a micro-balance enclosed in a glass tube from which the air could be evacuated. There are two balls attached to the ends of a rod that

---

[5]Thallium is a chemical element with atomic number 81.

[6]Philosophical Transactions of the Royal Society, Vol 164, 1874, p501-527.

is balanced in the middle, on a pivot. He tried many different types of balls and rods, until settling on balls made of pith [7] and a rod made of straw. These were the lightest and most responsive materials he could find. The pivot is a very thin metal rod, slightly shorter than the diameter of the glass tube. It is sharpened on both ends to a needle point and simply rests on the inside wall of the tube.

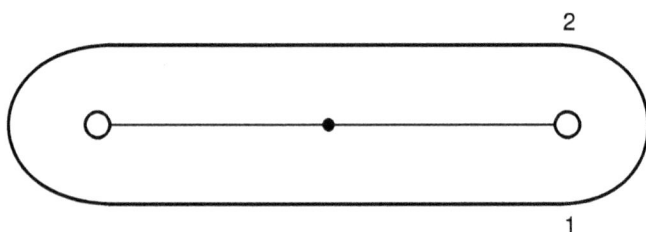

Figure 6: Micro-balance.

Starting with atmospheric air pressure in the tube, he placed a heat source at position 1 in the figure and watched the ball above it rise up. Placing the heat source at position 2 also made the ball rise but by much less. This was expected, since the air currents created by the heat, would tend to lift the ball.

Next he began evacuating air from the tube in a series of steps. At each step he once again looked at the effect of the heat source on the ball. The ball would continue to rise as the pressure went down but less and less vigorously. When he got down to a pressure of about 12 mmHg[8]

---

[7]Pith is a soft spongy material found in the stems of plants.

[8]The unit mmHg is millimeters of mercury. See the "Physics Reference" appendix for definitions of pressure units.

(1600 Pa) inside the tube, the ball would scarcely move at all and by the time he got to 7 mmHg (933 Pa), no source of heat could get the ball to budge.[9]

It was clear to him that at this low pressure there was not enough air for thermal currents to lift the ball. A lighter ball would still move up, but for any ball, no matter how light, movement would stop on approach to a total vacuum. This is true if thermal air currents were the only thing making the ball move and that would be the end of the story. But Crookes had enough curiosity and imagination to continue evacuating the air past this point.

His inquisitiveness was soon rewarded when he found that he could once again get the ball to move up. At a pressure of 3 mmHg (400 Pa), 4 mm below the point where the ball would no longer budge, he could get it to move up with the same force as it did at atmospheric pressure. Continuing to lower the pressure, he found that the ball would move up with even more force and, surprisingly, with much less heat. Just placing his finger under the ball would get it to shoot up. He also found that placing the heat source above the ball would make it go down whereas at higher pressures it made it go up. Clearly there was something different going on here, at these lower pressures.

Crookes continues on with an almost bewildering number of additional experiments. First, in the place of heat, he tries using cold, in the form of a lump of ice. He gets almost the same results but in the opposite direction. The

---

[9]When Crookes did these experiments, devices for measuring low pressures were very crude so these pressure measurements are probably not very accurate.

ball always moves toward the cold. He then tries countless combinations of different materials and experimental arrangements in an attempt to unambiguously identify what is making the balls move. He eventually concludes that it must be radiant energy that is making them move, i.e. radiation pressure.

The most striking thing about his conclusions at the end of the paper is what he says about the possibility that it is the residual gas in the tube that may be causing the movement. He says:

"In favour of this explanation it may be urged that a highly rarefied gas may be much more mobile than when it is denser, and therefore the more rapid impingement of its particles, when set in ascension by warmth, would increase their mechanical action. Increased momentum may counterbalance diminished number."

This is interesting because it comes close to describing what actually does make the balls move. But he dismisses the idea by saying that:

"it is most difficult to believe that the residual air ... can exert, when gently warmed by the finger, an upward force capable of instantly overcoming the inertia of a mass of matter weighing several grains[10], and setting it in motion."

The clincher for him was the observation that movement of the balls stopped when the pressure reached 7 mmHg but started up again when the pressure got lower. He

---

[10]A grain is a unit of weight equivalent to approximately 65 milligrams. There are exactly 7000 grains in one pound (0.45 kg).

reasoned that if the air could not move the balls at 7 mmHg then surely it couldn't do so when there was less of it. What he didn't realize at the time is the fact that a gas does behave differently at low pressures.

The density of molecules in a gas is proportional to the pressure and when the density gets low enough, the number of collisions between molecules becomes negligible. This means it takes much longer to establish thermal equilibrium. In other words, the heat does not spread easily among the molecules. The molecules essentially only interact with surfaces inside the vacuum chamber. They can then pick up energy from a hot part of the inside wall and directly transfer it to one of the balls. This happens at extremely low pressures. But Crookes also observed the effect at pressures where this explanation does not hold. In that case there is something more subtle going on which we will discuss later in the book.

One of the things that led him astray was his apparent conclusion that the force on the ball would continue to increase as the pressure dropped and reach a maximum at zero pressure. At that point, the only thing that could possibly account for the force would be the radiation pressure.

He drew this conclusion because of the limitations of his equipment. Had he been able to continue lowering the pressure to hundredths of a mmHg he would have seen the force on the ball peak and then start to decrease to zero. Then it would be very clear that the residual gas molecules were responsible for the force on the ball and the role of the heat or light source was simply to provide the molecules with energy.

At the very end of the paper he still holds on to the idea that the effect may somehow be connected with the force of gravity and may even be the key to solving "some as yet unsolved problems in celestial mechanics.". He concludes the paper with the following statement:

"Although the force of which I have spoken is clearly not gravity solely as we know it, it is attraction developed from chemical activity, and connecting that greatest and most mysterious of all natural forces, action at a distance, with the more intelligible acts of matter. In the radiant molecular energy of solar masses may at last be found that 'agent acting constantly according to certain laws' which Newton held to be the cause of gravity."

In discussions about the history of the radiometer it is often said that Maxwell[11] was somehow overly eager to believe that Crookes had discovered evidence for the existence of radiation pressure, a phenomenon that Maxwell had predicted when formulating the laws of electromagnetics. But the evidence that Crookes presents in this first paper in no way rules out the possibility that the force is due to radiation pressure.

Crookes had not yet constructed the more familiar modern form of the radiometer with vanes colored white on one side and black on the other. These vanes turn contrary to the fact that radiation pressure should exert twice as

---

[11] James Clerk Maxwell (1831 – 1879) was a Scottish physicist who made contributions to electromagnetics, thermodynamics, and kinetic theory.

much force on the white side as on the black side. This is strong evidence that radiation pressure is not at work. But it is not clear if Maxwell had seen one of these radiometers when he first thought that Crookes may have found evidence for radiation pressure. If all Maxwell had to go on was the evidence in Crookes' first paper then his conclusions were quite understandable.

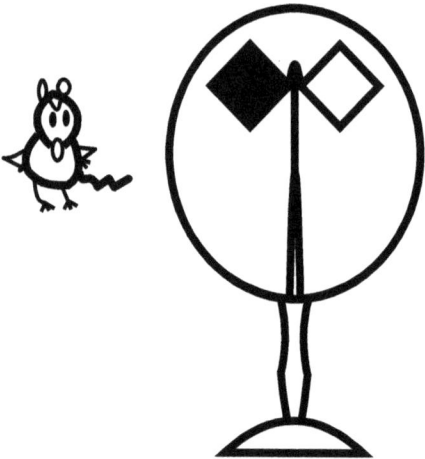

In his second paper published in 1875 [12] Crookes describes a device that's starting to look a little more like the radiometer most people are familiar with. Figure 7 is an example of what it looked like. There is a thin glass bar suspended horizontally from a silk thread inside a glass bulb. On each end of the bar there is a ball or disk made of various substances. He notes that the disks must have large surface area and low weight to get the highest sensitivity. He used disks made of pith, wings of butterflies and dragonflies, dried and pressed rose leaves, thin mica and selenite, and iridescent films of blown glass.

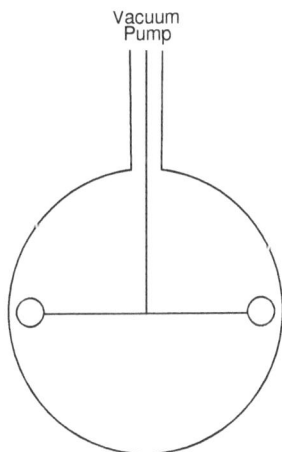

Figure 7: Early radiometer.

This device was so sensitive that touching the outside of

[12]On repulsion resulting from radiation - Part II, Philosophical Transactions of the Royal Society, Vol 165, 1875, p519-547

the bulb near one end of the bar with a finger would turn the bar around 90 degrees. Moving a piece of ice around the outside of the bulb would cause the bar to follow it the way a compass needle follows a magnet. Crookes describes how bringing a candle flame near the globe would cause the bar to oscillate continuously. When demonstrating the device at a meeting of the Royal Society, it is this behavior that gave Osborne Reynolds the clue that it was not light radiation that was directly causing the movement inside the bulb but rather the dynamics of the residual gas.

Reynolds published a paper[13] where he suggested that it was evaporation and condensation of molecules on the balls at the ends of the rod that was causing them to move. When a ball was exposed to heat (infrared radiation), the molecules on its surface would evaporate and, by conservation of momentum, cause the ball to move in the opposite direction, away from the heat source. If the gas on one side of the ball was cooled it would cause molecules to condense onto the ball, causing the ball to move in the direction of the cooled gas.

Crookes rejects this theory and describes experiments where he heats the bulb "to a dull red heat" while continuing to exhaust it for two days. Afterwards the behavior was the same. The ball was repulsed by heat and attracted by cold. He concludes that "It is impossible to conceive that in these experiments sufficient condensable gas or vapour was present to produce the effects Professor Os-

---

[13]On the forces caused by evaporation from, and condensation at, a surface, Proceedings of the Royal Society, Vol 22, Jun 18 1874, p401-407

borne Reynolds ascribes to it."

It is also in this second paper that we see a connection between this work and Crookes's interest in spiritualism. He mentions some work by a Baron von Reichenbach[14] that purportedly shows "that there is a peculiar emanation or aura proceeding from the human hand". If this is true, Crookes says, then it is not inconceivable that it could have an effect on the operation of the radiometer. To test this, he tried numerous experiments to, in his words, "see if there was any difference in action between the fingers and a tube of water of the same temperature." He could find none. He also looked to see if there were any anomalous effects due to crystals and chemical reactions that could not be explained by heat. None were found.

Crookes concludes the paper by saying that he wanted to avoid advancing a theory until he had more facts, since in his words, "The facts will then tell their own tale; the conditions under which they invariably occur will give the laws; and the theory will follow without much difficulty." He seems to be having some doubt at this point as to whether or not radiation pressure is causing the motion.

He ends the paper with a quote from Sir Humphry Davy [15] that reads almost as an apology for being more of an experimentalist than a theorist.

"When I consider the variety of theories which may be

---

[14]Karl Ludwig von Reichenbach (1788-1869) was a German chemist who believed that all living things emit a form of energy that he called the Odic force.
[15]Sir Humphry Davy (1778-1829) was an English chemist and inventor.

formed on the slender foundation of one or two facts, I am convinced that it is the business of the true philosopher to avoid them altogether. It is more laborious to accumulate facts than to reason concerning them; but one good experiment is of more value than the ingenuity of a brain like Newton's."

In Crookes' third and fourth papers, published together in 1876, [16] he unveils a new device that allows him to get continuous rotation. He does this by replacing the silk thread with a low friction pivot. The device is now starting to resemble the familiar Crookes radiometer. An example of what it looked like is shown in figure 8.

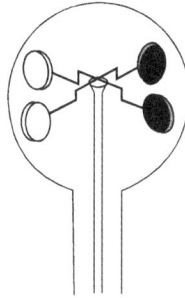

Figure 8:   First radiometer.

The balls have been replaced by disks and instead of just two, there are four. The sides of the disks are alternately blackened with lampblack [17]. The new device also has a new name. He calls it a *radiometer* or *light-mill* because, as he says, "it serves to measure the amount of radiation falling upon it by the velocity with which it revolves".

The paper goes on to detail many of his experiments with this new radiometer. He tried adding more disks, up to

---

[16]On repulsion resulting from radiation - Parts III & IV, Philosophical Transactions of the Royal Society, Vol 166, 1876, p325-376.

[17]This is basically soot from some combustion process such as a burning candle or oil lamp.

ten, but settles on four as the most practical number. Different colors for the disks were tried but simple lampblack on one side and white pith on the other gave the best results. He looked at the relationship between the number of candles illuminating the radiometer and the speed of rotation. The effect of illumination direction was also tested. There was faster rotation when the illumination was uniform in all directions, as opposed to a single direction only.

He tests the effect of boiling water and ice near the radiometer. He puts a magnet on 2 arms of a 10 arm radiometer, then suspends a magnet nearby which oscillates according to the rotational speed of the radiometer. He created some radiometers with disks of glass, and another with a mirror in it, for the purpose of reflecting the radiometer's motion onto a wall or ceiling for demonstrations. He recorded the effect of different parts of the solar spectrum on a specially designed radiometer. He coated the pith disks of the radiometer with various substances and noted the effectiveness each had on the turning of the radiometer.

One of the more interesting observations made in this paper is that, if the radiometer is placed in a uniformly heated space, the turning of the vanes[18] will eventually stop, once the radiometer's temperature reaches the temperature of the surroundings. On the other hand, if it is placed in a uniformly lighted space, it will continue to turn as long as there is light.

---

[18]From here on we will refer to the radiometer arms and what is attached to them as vanes.

We now know that the radiometer will only turn if there is a temperature difference between the two sides of each vane. For visible light the two sides never reach the same temperature. The black side absorbs more radiation and will always be warmer than the white side. For infrared radiation (heat) things are a little more complicated. The glass bulb surrounding the radiometer absorbs infrared radiation and heats up. This heat is then conducted by the residual gas molecules to both sides of the vanes, eventually heating them to the same temperature. At this point the turning stops.

In a postscript to the paper, written almost a year after his experiments, he concludes that the movement of the radiometer vanes is not due to radiation pressure after all but to the dynamics of the residual gas. He credits George Johnstone Stoney (1826-1911), an Irish physicist and Fellow of the Royal Society, with first coming up with this explanation. He states that the rotation of the radiometer vanes is a maximum at 50 millionths of an atmosphere[19] (5 Pa). The fact that there is a maximum and that the rotation doesn't just continue to increase as you go down in pressure indicates that the residual gas plays a major role in making the vanes turn. An experiment by Arthur Schuster (1851-1934), a German born British physicist and Fellow of the Royal Society, in 1876 provided further proof that the residual gas is making the vanes turn and not the radiation pressure. We will discuss this experiment in detail later in the book.

---

[19]Devices for measuring very low pressures during Crookes' time were not very accurate so this is an approximate value.

Figure 9: George Johnstone Stoney was credited by Crookes as first to explain the movement of the radiometer in terms of dynamics of the residual gas. Public domain image courtesy of Wikimedia Commons at https://commons.wikimedia.org/wiki/ File:GeorgeJohnstoneStoney(1826- 1911),Undated(DateGuessedEarly1890s).jpg

Figure 10: Arthur Schuster showed that the force rotating the vanes of the radiometer is generated internally. Public domain image courtesy of Wikimedia Commons at https://commons.wikimedia.org/wiki/ File:Arthur_Schuster.jpg

In the postscript he also, more or less correctly, describes the effect of the mean free path[20] on the gas dynamics and the different effects of infrared and visible light. The glass bulb absorbs infrared light and heats up while it passes the visible light which then heats up the vanes. This explains why the radiometer behaves differently when exposed to uniform light and uniform heat.

In Crookes' fifth paper, published in 1878,[21] Crookes continues his radiometer investigations but now with the knowledge that it is the gas dynamics and not the radiation itself which is making the vanes move. He tries coating the radiometer vanes with a variety of substances to see which will produce the greatest force on the vanes. He tries making vanes out of different materials such as mica, and metal, and tries various shapes from round to square to cup and cone shaped vanes.

One of the major discoveries in this paper is that the shape and orientation of the vanes has a big effect on their performance. Crookes put together a radiometer with vanes made of gold foil blackened on one side and tried it using the light from a candle. As usual he found that the blackened side of the vanes were repelled by the light. What was unusual was that the unblackened side of one of the vanes was also repelled. Looking at this vane more closely he noticed that it was crumpled and slightly curved at the edge near the container wall. This prompted him

---

[20]The mean free path is the average distance a gas molecule travels before colliding with another molecule.

[21]The Bakerian Lecture - On repulsion resulting from radiation - Part V, Philosophical Transactions of the Royal Society, Vol 169, 1878, p243-318.

to do a series of experiments where he purposely turned up the outside edge of the vanes by varying degrees to see what effect it had.

He got the biggest result with the vanes turned as shown in figure 11. Here the vanes are turned at 45 degrees away from the radial lines so that they more directly face the wall of the chamber. He colored the sides in two different ways. With sides B black and sides A reflective there was weak rotation in the direction of the arrows. With sides A black and sides B reflective there was strong rotation in the direction of the arrows. He says that with sides A black and vanes made out of mica the radiometer was the most sensitive to light than any he had yet constructed. What this experiment seems to show is that the walls of the chamber play a major role in making the vanes turn.

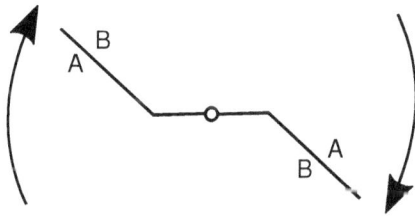

Figure 11:   Vanes turned 45 degrees.

He called these radiometers "Sloping Vaned Radiometers" and built a series of them with four arms as shown in figure 12. Both sides of all the vanes in these radiometers were the same color. They were either all reflective or all blackened. In the usual radiometer, where the vanes were oriented straight along radial lines, this would prevent any turning. In this case however, they did turn, when exposed to the light of a candle. They turned in the direction of the arrows shown in the figure but the rotation did not start immediately upon exposure to the candle. This suggests that the glass of the bulb walls had to heat up to provide the rotation energy. Of all these radiometers, the one with all the vanes blackened turned the fastest. In this case the visible light of the candle raised vane temperatures and provided extra rotation energy.

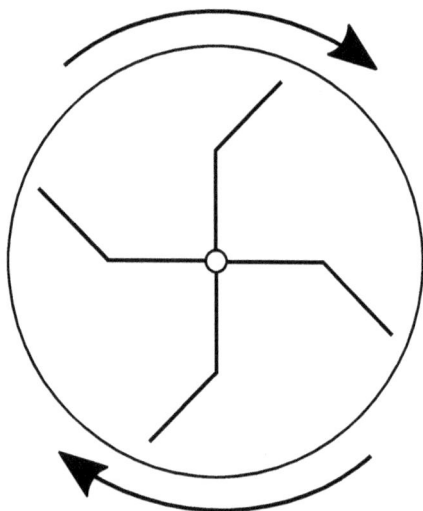

Figure 12: Sloping vaned radiometer.

Further experiments with these radiometers showed that, if heat was applied around the circumference of the bulb, it made the vanes turn in the direction of the arrows in the figure. If it was applied on the top or bottom of the bulb, it made the vanes turn in the opposite direction. He explains this as follows. When heat is applied around the circumference it creates a "molecular disturbance, which presses towards the center and strikes the sloping vanes, driving them round as if a wind were blowing on them."

When heat is applied to the top or bottom he says that the "molecular pressure" from those areas "strike the inner surface of the sloping vanes" driving them in the opposite direction.

He next looks at the performance of cone shaped vanes made of aluminum, as shown in figure 13. When exposed to candle light they turned in the direction of the arrows at a rate of about 20 revolutions per minute. When the light was blocked from hitting the convex part of the cone the turn rate was only 10 rpm. When it was blocked from hitting the concave part the turn rate also dropped to 10 rpm. Doing the same experiment with cones made of mica showed very little rotation.

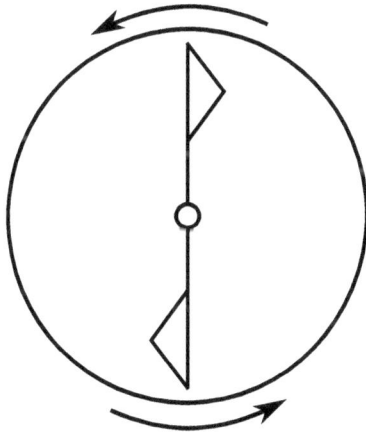

Figure 13:   Cone vaned radiometer.

From these experiments it was clear to him that the orientation of the vanes with respect to the walls was very important. He concluded that it was in fact more important than having a difference in vane colors. In an effort to find the optimal vane shape he next tests cylindrically shaped vanes as shown in figure 14. The radius of curvature for vanes A, B, C, and D where respectively 5, 10, 20, and 30 mm. All four vanes were 10 mm high and 10 mm across the ends of the arc. He tested them all at the same pressure and same illumination. The number of revolutions per minute for vanes A, B, C, and D respectively were 21.4, 15, 10.3, and 4.6. The vanes which were more deeply curved away from the radial line experienced faster rotation which agreed with his cone and sloped vane radiometer experiments.

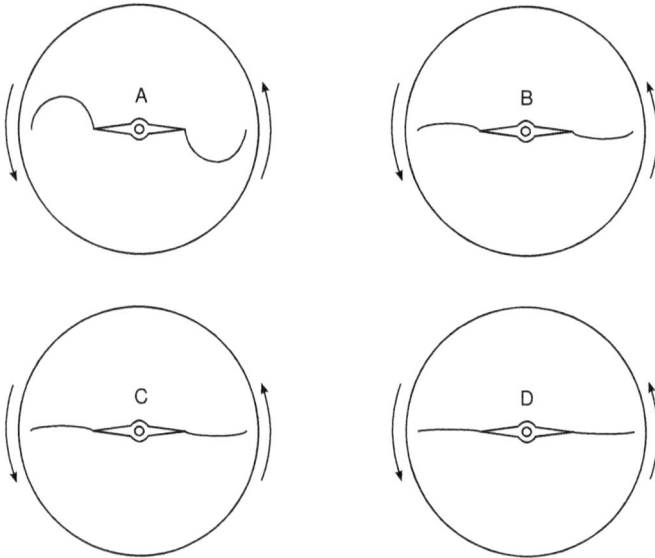

Figure 14: Cylindrical vaned radiometers.

He goes on to test spherical cup shaped vanes and gets the same result. The more deeply curved vanes rotated faster. He experimented with gold and aluminum cups as vanes, blackening alternately the concave and convex sides, no sides, or one or the other. He found the highest rotation rate using aluminum cups blackened on both sides. Using a radiometer with aluminum cups as vanes and both sides blackened, he illuminates it with candles while varying the pressure from 577 millionths of an atmosphere (58 Pa) down to 0.2 (0.02 Pa). At each pressure he measured the revolutions per minute. The maximum rotation speed occurred at 41.5 millionths of an atmosphere (4.2 Pa). We will take a closer look at this data in the chapter on how the radiometer works.

Crookes was aware that glass blocks infrared radiation so he devised an experiment in which the infrared source was inside the bulb. The experiment is shown in figure 15. The infrared source consists of a wire loop that can be heated red hot by passing an electric current through it. Four thin clear mica vanes sit about 5 millimeters above the loop and are free to rotate in a horizontal plane parallel to the plane of the loop as shown in the figure. Each vane is at an angle of 45 degrees with respect to the horizontal so that when looked at from the side it has the appearance of a forward slash /.

When the wire was heated the vanes would rotate rapidly such that the slash appeared to move from right to left. As long as the wire was kept hot the rotation continued unabated. At a pressure of 34 millionths of an atmosphere (3.4 Pa) the vanes rotated at 200 rpm. Lowering the pressure to 3 millionths of an atmosphere (0.3 Pa) took the rotation up to 300 rpm and it stayed there down to 1 millionth (0.1 Pa). He could not lower the pressure further with this setup. The interesting thing is that with a regular radiometer the rotation speed would go down

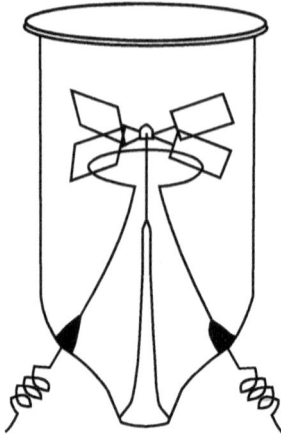

Figure 15: Radiometer with infrared source inside.

when the pressure went below 34 millionths of an atmosphere where as in this case it went up. Putting the heat source inside the bulb changed the dynamics significantly.

In Crookes' sixth and final paper in this series, submitted in 1878, and published in 1879[22], he begins with experiments where he tries to understand the distribution of what he calls the molecular lines of pressure inside the radiometer. These are hypothetical lines along which the motion of the molecules cause the forces that move the vanes. He does this by placing mica sheets in various positions relative to the vanes.

One of the findings from these experiments is that in a radiometer with plain cup shaped vanes (no blackened surfaces), both convex and concave sides alone are able to turn the arms but the effect of the convex side is about 10 times greater than the concave side. Overall this makes the arms turn like a cup anemometer in reverse.

Another interesting experiment described in this paper is shown in figure 16. This is a radiometer with a single arm that has a hollow mica prism attached to it with a counter weight. It rotates above a blackened mica disk that has a platinum wire loop underneath which can be heated by passing an electric current through it, similar to the setup shown in figure 15.

When the pressure in the radiometer is reduced[23] the arm does not turn but as soon as the wire under the disk is

---

[22]On repulsion resulting from radiation - Part VI, Philosophical Transactions of the Royal Society, Vol 170, 1879, p87-134.

[23]As in many of his experiments Crookes leaves out details such as what exactly the pressure was.

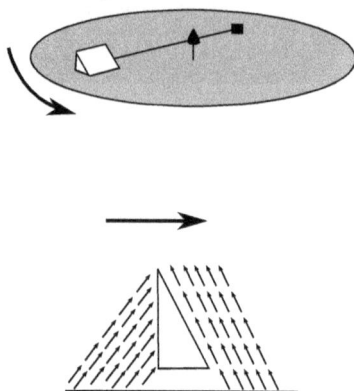

Figure 16: Mica prism radiometer.

heated it starts to turn and continues turning as long as the wire is hot. He concludes from this that the molecular lines of pressure are not strictly normal to the surface but come off at an angle, driving the prism forward as shown in the figure.

Crookes next describes what he calls a turbine radiometer. This radiometer, shown in figure 17, can be optimized for light shining from above or below. The design allows for a higher density of vanes arranged in a windmill type pattern. If the top of the disks are blackened then it is more sensitive when illuminated from above. If the bottom of the disks are blackened then it is more sensitive when illuminated from below.

If the turbine radiometer is floated in a vessel of ice cold water, and exposed to the air of a warm room above, it rotates rapidly, acting as a heat engine. By floating the radiometer in hot water instead and exposing it to cold air above, it revolves in the opposite direction.

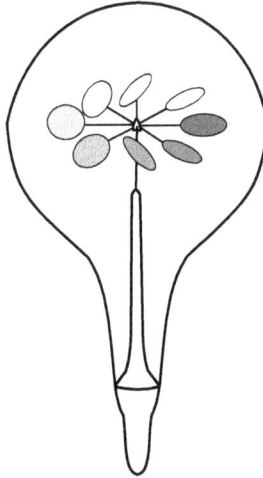

Figure 17: Turbine radiometer.

Crookes suspected that what caused the movement of the radiometer arms was pressure between the vanes and the inner surface of the bulb. This is what he calls molecular lines of pressure. He illustrates these lines of pressure as straight arrows acting between surfaces. The straight arrows are ostensibly the free paths of the molecules or possibly some sort of average movement of the molecules as in a pressure or sound wave. His exact conception of what these lines are is never made exactly clear.

It did seem apparent to him that the shorter the distance over which these lines had to act, the stronger the pressure they could exert on the surfaces. To test this hypothesis he constructed a radiometer with a small and large bulb joined together. The same set of arms could be set to rotate in either bulb. In the small bulb the vanes on the arms rotated very close to the bulb surface while in the large bulb they were further away. He found that in the small bulb the arms did rotate about 50 percent faster than in the large bulb, thus verifying his hypothesis.

Seeing that the glass bulb played a role in exerting pressure on the arms and making them turn, he decided to look at the effect of altering the inside surface of the bulb. He placed an aluminum band around the middle of the inside surface of the bulb. The width of the band was slightly larger than the width of the vanes and the inside surface of the band was blackened. At the center of the bulb he had the standard four arms with vanes blackened on one side. He found that the arms rotated significantly faster with the aluminum band than without it.

In 1876 the physicist Arthur Schuster did an experiment where he suspended a radiometer from 2 thin wires in the manner of a bifilar pendulum. When the radiometer was illuminated, the vanes turned one way and the suspended bulb turned the opposite way. It's not clear if Crookes was inspired by this experiment or not, but he performed a similar experiment where the bulb of the radiometer turned while the vanes were stationary. He did this by first placing a magnetic needle on the arm of a radiometer. He then floated the entire radiometer in water and illuminated it. The arms turned as usual but when a powerful magnet was used to stop the arms, the radiometer bulb started turning in the opposite direction. This clearly showed that a force opposite to that on the vanes was being exerted on the bulb, in accordance with Newton's third law.

In another experiment, Crookes placed mica sheets on the inside wall of a radiometer as shown in figure 18.

There are two parallel mica sheets about a millimeter apart, mounted on the inside wall of the bulb. One of the sheets is clear, the other is blackened on the side facing away from the clear sheet. The vanes are made of clear mica and they pass through a hole in the sheets with a clearance of about a millimeter. Shining light on only the vanes or the clear mica sheet produces no rotation. When the mica sheet with the blackened side is illuminated, Crookes notes that the arms rotate rapidly, as if "blown round by a wind issuing from the black surface". The rotation continues as long as the light is on. In this case it is clear that the molecular motion making the arms turn originates on the blackened mica sheet and not on the vanes.

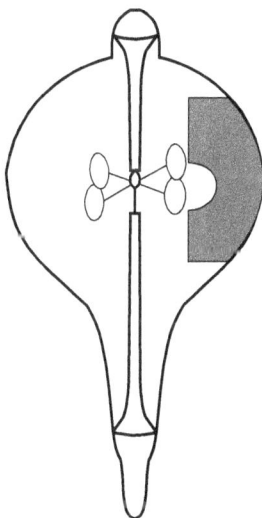

Figure 18: Radiometer with mica sheets fixed on inside wall.

One physicist who was very interested in Crookes' work,

and who himself did research on the radiometer, was Sir George Stokes. Stokes was a secretary of the Royal Society and one of the editors of the "Philosophical Transactions of the Royal Society," where Crookes published much of his work. Stokes often suggested experiments to Crookes, as well as theoretical explanations for his results. On the suggestion of Stokes, Crookes created the unique radiometer shown in figure 20.

This was your standard four arm radiometer but the vanes were made of polished aluminum with both sides identical, i.e. no blackening on either side. There are three thin clear mica sheets fixed to the inside of the bulb as shown in the figure. They are oriented away from the axis of rotation. The vanes pass through notches in the sheets as shown in the figure.

When illuminated from points a, b and c the arms rotate rapidly in the direction of the arrow. Crookes writes that breathing gently on the bulb will make it rotate in a direction opposite to the arrow. This is presumably with the illumination at a, b and c removed. Likewise, placing a hot glass cover over the entire radiometer will make it rotate opposite to the arrow until the glass cools, at which point it will rotate in the direction of the arrow. Crookes writes that: "the strongest action is produced by warming the bulb."

In the most common radiometer design, the energy that causes the arms to turn originates on the blackened side of the vanes. Crookes likens these blackened surfaces to the high temperature reservoir of a heat engine, with the glass bulb being the low temperature reservoir. The vanes

Figure 19: Professor Sir George Stokes, circa 1860's, who helped Crookes to understand the radiometer and made experimental suggestions to Crookes. Public domain image courtesy of Wikimedia Commons at https://commons.wikimedia.org/wiki/File:Ggstokes.jpg

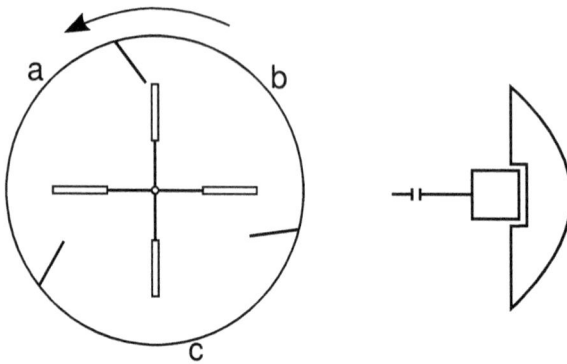

Figure 20: Stokes radiometer.

are a poor reservoir however, due to their small size. They cannot be heavy since this would increase the friction in the pivot and they cannot be too large since this would increase drag. They also have to be poor heat conductors so that a temperature difference between the two sides can be maintained.

To overcome these limitations, Crookes figures that the high temperature reservoir, or heater as he calls it, should be stationary. Then there will then be no limits on how large and heavy it can be. In his view the heater, "acts as if a molecular wind were blowing from it, principally in a direction normal to the surface." With this in mind, the vanes, or moving part of the radiometer, should be designed to take maximum advantage of the wind, to produce motion. He decides to call this new radiometer design an otheoscope (from the Greek meaning, I propel).

Figure 21 shows two otheoscope designs. In design A, there is a fixed horizontal mica disk blackened on the upper side. Above it are four vanes of polished aluminum on both sides, set at an angle of 45 degrees with respect to the disk. The vanes rotate as close as possible to the disk. They begin to rotate as soon as light shines on the blackened side of the mica disk. In design B, the mica disk has been replaced with a copper disk blackened on the upper side. The vanes are now made of mica and are arranged in the manner of fan blades, with as many packed together as possible. Design B was found to be more sensitive than design A. In another version, Crookes used aluminum cups as vanes oriented at 45 degrees with the concave side facing down. This also turned out to be very sensitive.

These designs were very sensitive to heat. If the top part of the bulb was heated, the vanes would turn opposite to the normal direction. If heated from below, the rotation is delayed until the disk below the vanes heats up. The vanes then turn in the normal direction. When the entire otheoscope is plunged into hot water the vanes turn opposite to the normal direction until everything is at a uniform temperature. When removed from the hot water, the vanes turn normally until everything cools down.

This is the last of Crookes' papers devoted solely to what he calls "repulsion resulting from radiation". In the process of investigating this, he created a bewildering number of different radiometers and variations on them, including the one most people are familiar with today. He was an indefatigable experimenter, on par with great experimentalists such as Thomas Edison, Nikola Tesla, and Michael Faraday.

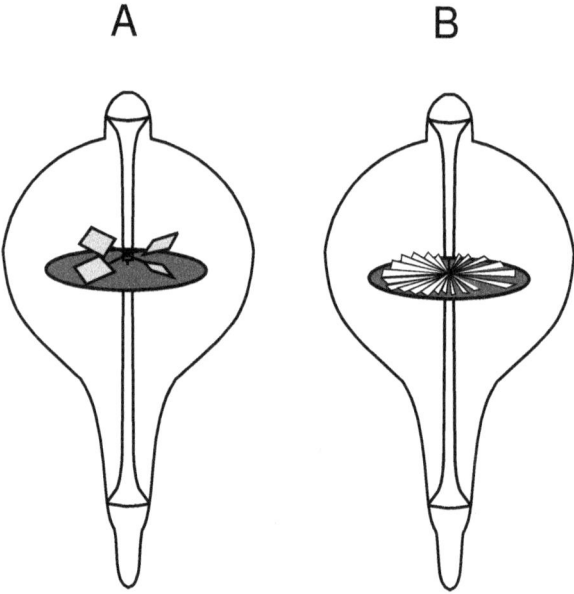

Figure 21: Otheoscope.

He did have the good fortune of working with a remarkably talented assistant named Charles H. Gimingham. Gimingham had a short life, dying at the age of 37. He worked for Crookes for 12 years, during which he improved the Sprengel vacuum pump. At the request of Crookes, Gimingham submitted a paper to the Royal Society on his improvement of the Sprengel pump, without which much of the radiometer research would not have been possible[24]. More information about the life of Charles Gimingham can be found here:

Charles and Edward Gimingham – Light Bulb Pioneers
https://ietarchivesblog.org/2017/05/10/charles-and-edward-gimingham-light-bulb-pioneers/

It is interesting to observe the evolution of Crookes' ideas about what caused the radiometric force. He went from thinking he had observed the effect of radiation pressure to thinking he had discovered a new state of matter. His idea seems to be have been that, when a gas becomes rarefied enough, it no longer behaves like a gas but enters into a new state, where interaction with a heated surface can convert the random motion of the molecules into a collective rectilinear motion in a particular direction. This motion then creates pressure on the surfaces it originates and terminates on.

In his later research he attempted to observe this motion by turning his radiometers into what is essentially now called a cathode ray tube. The next paper he published

---

[24]On a new form of the 'Sprengel' air-pump and vacuum-tap, Proceedings of the Royal Society of London, Vol 25, Dec 7 1876, p396–402

in Philosophical Transactions of the Royal Society was titled "On the illumination of lines of molecular pressure,
and the trajectory of molecules". In it he tries to draw
connections between cathode rays and his lines of molecular pressure. This research led him to the development
of the Crookes tube which is another interesting story
but here is where our story stops. For more information
about Crookes, see the appendix for a timeline of his life
and references listed in the bibliography.

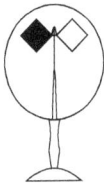

In this chapter we will dive a little deeper into the question of how the radiometer works. What is it that makes the vanes turn?

Radiation Pressure

Let's first look at whether radiation pressure can play any role in causing the rotation. We'll use the sun as a radiation source and assume, ideally, that we have a solar irradiance[25] of one kilowatt per square meter $I = 1000$ W/m^2. This is about as strong as sunlight ever gets on the Earth's surface. What effect does light of this intensity have when it falls on a radiometer arm with the pair of vanes shown in figure 22? Here we have two flat circular vanes, one colored white, the other black. The arm connecting the vanes is free to rotate about a vertical axis at its center.

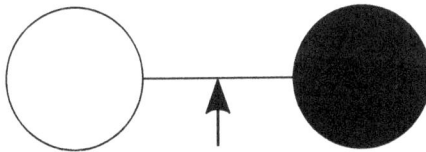

Figure 22: Radiation pressure.

Assume the white vane is a perfect reflector, meaning it reflects all light that falls on it. Assume the black vane

---

[25]Irradiance is the amount of power per unit area received by a surface.

is a perfect absorber, meaning it absorbs all light that falls on it. If the light waves are perpendicular to the plane containing the vanes then the pressure due to total absorption is the irradiance divided by the speed of light, $P = I/c$, while the pressure due to total reflection is twice this amount. This means the white vane has twice as much pressure on it as the black vane. So with sunlight incident normal to the page and with no other forces at work, the arm would turn such that the black vane comes out of the page and the white vane goes into the page. Looking at it from above, down the axis of rotation, the arm would rotate clockwise.

This is not however the way this arm would actually turn inside a radiometer. The black vane would go into the page, away from the light source, and the white vane would come out of the page, toward the light source. Looking from above, down the axis of rotation, the arm would rotate counterclockwise. This fact alone indicates that radiation pressure is not what is making the arm turn. In a radiometer, the radiation pressure certainly does exist but it acts opposite to the way the arms turn, i.e. it is impeding the turning of the arms.

By how much is it impeded? Calculating the pressure, $P = I/c$, we get

$$P = \frac{1000 \text{ W/m}^2}{3 \cdot 10^8 \text{ m/s}} \approx 3 \cdot 10^{-6} \frac{\text{N}}{\text{m}^2}$$

This is an extremely small pressure. Now if the area of each vane is one square centimeter then the net force on a vane is on the order of a nanonewton. To put this in perspective, such a force is equivalent to the weight of about 100 typical bacterial cells on the Earth's surface. A force this small is negligible. The frictional force on the pivot will usually be much greater than this.

Bifilar Pendulum

There are other ways to show that radiation pressure is not causing the rotation. One way is to make a bifilar pendulum out of the radiometer. You do this by suspending it with two thin wires or threads as shown in figure 23. When the radiometer is exposed to light, the vanes will turn as usual but the glass bulb will also turn. How it turns with respect to the vanes indicates if the force is internal to the bulb or external. If the bulb turns in the same direction as the vanes, the force is external, which in this case means it has to be due to radiation pressure. If it turns opposite to the vanes, the force is internal, meaning there is something inside the bulb, such as residual gas, causing the rotation.

This experiment was first done in 1876 by Arthur Schuster, who was a colleague of Osborne Reynolds. The results provided definitive proof that radiation pressure is not the cause of rotation, i.e. the vanes and bulb turned in opposite directions.

The basis for the experiment is Newton's third law of motion: for every action there is an equal and opposite reaction. If the force that causes the turning is due to radiation pressure then, with the radiometer suspended as a bifilar pendulum, you would see both the vanes and

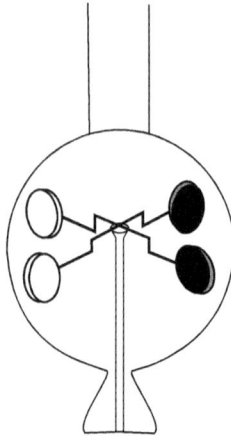

Figure 23: Bifilar radiometer.

bulb turn in the same direction. The frictional force in the pivot, on which the vanes rotate, would drag the bulb in the same direction as the vanes.

On the contrary, if the force is generated internally, by the residual gas molecules, then there would be a reaction on the walls of the glass bulb surrounding the gas. This would cause the vanes and the bulb to turn in opposite directions. If you do the experiment this is exactly what you see.

When the radiometer is exposed to light, the vanes begin to rotate and the reaction force turns the bulb in the opposite direction. The bulb quickly reaches a maximum rotation angle and then begins a decaying oscillation about its initial position, once the vanes have reached a constant steady state speed. The bulb eventually comes to a stop at its initial position.

Drag Forces

The fact that the vanes eventually reach a constant steady state speed is another important clue to explaining the radiometer's behavior. What it means is that there must be a velocity dependent force that opposes the force making the vanes turn. This is a drag force due to the residual gas in the bulb. So the total force on the vanes can be written as

$$F_T = F - f(v)$$

where F is the force causing the vanes to turn and $f(v)$ is the drag force. The drag force is usually a linear function of the velocity, v, of the vanes, so we can write it as $f(v) = a_1 v$ where a_1 is the linear drag coefficient. At very high velocities $f(v)$ may switch over to having primarily a quadratic dependence on the velocity, $f(v) = a_2 v^2$ where a_2 is the quadratic drag coefficient.

These relationships are only approximate and coefficient values can change as the velocity changes but such details are not important for this experiment. What is important is that, as the rotation velocity increases, the drag force increases until $f(v) = F$, at which point the total force on the vanes is zero. The reaction force on the bulb is then also zero, so the restoring force of the bifilar pendulum suspension will cause a decaying oscillation of the bulb back to its initial position.

Now we may ask, with the vanes rotating at constant speed and the bulb settled back to its initial position, what happens when the light is suddenly cut off? The total force on the vanes is now just the drag force, $F_T = -f(v)$ so they will slowly come to rest. The force on the bulb is the negative of this, meaning it is initially deflected in the same direction as the vanes are turning. After reaching a maximum deflection angle it experiences a decaying oscillation back to its initial position.

Estimating the Forces

With careful measurement it is possible to use this experiment to extract semi-quantitative information about the dynamics. By measuring the maximum deflection angle of the bulb, when the radiometer is exposed to light, you can get an estimate of the force making the vanes turn. If θ is the maximum deflection angle then the force on the vanes is given by

$$ F = \frac{d^2}{hl} w\theta $$

where w is the weight of the radiometer, d is the distance between the filaments on which the radiometer is suspended, h is the length of the filaments, and l is the distance from the pivot to the center of a vane. For a derivation of this formula see the problem section.

By measuring the angular velocity of the vanes and how long it takes them to stop when the light is suddenly blocked, you can get estimates for the drag force and the frictional force in the pivot. If for example there is only a frictional force in the pivot and no drag force then the time it takes for the vanes to stop turning is

$$t = \frac{\omega_0 I}{\tau}$$

where ω_0 is the initial angular velocity, I is the moment of inertia of the vanes and τ is the torque due to the friction in the pivot. For a derivation of this formula see the problem section. Even though we know there must be a drag force in addition to the friction, we can still solve this equation for τ to get an order of magnitude estimate for the friction in the pivot.

With friction in the pivot and a linear drag force, the time it takes for the vanes to stop turning is

$$t = \frac{I}{a_1} \ln \left(1 + \frac{a_1}{\tau}\omega_0\right)$$

where a_1 is the linear drag coefficient. For a derivation of this formula see the problem section. Using the estimate for τ this equation can be used to get an estimate for a_1.

Convection Currents

Since we know now that it is the residual gas that makes the vanes turn let's look at whether convection currents in the gas could be at least partially responsible. From classical fluid dynamics, it is known that convection currents can be set up in a gas whenever there are temperature differences and an external force such as gravity. We certainly have these ingredients in the radiometer [26] but the evidence shows that the gas density is too low for convection currents to play much of a role in the dynamics. Crookes himself discovered this in his initial experiments.

Starting at atmospheric pressure he could make pith balls on a balance in a glass container move up by placing a heat source underneath them. This movement was obviously caused by the convection currents generated by the heat source. As he lowered the pressure however, the movement diminished and eventually stopped. At this point the convection currents either ceased to exist or were too weak to move the balls.

But as he continued to lower the pressure, he found that the balls suddenly started to move again. Clearly there was a change in the mechanism causing the movement and it occurred only when the pressure, and hence the density of the gas, was low enough. This mechanism, whatever it is, generates what we will call the radiometric force or pressure, to make it clear that it has nothing to do with convection currents.

---

[26]To our knowledge a radiometer has never been operated in a zero gravity environment and it would be interesting to see if it causes any change in the dynamics.

Radiometric Force as a Function of Pressure

In Crookes' later experiments he discovered that the radiometric force continued to increase as the pressure was lowered until it reached a maximum value and then started to decline, with the trend showing that it would go to zero at zero pressure. Experiments by many others have confirmed this. See for example the papers by Wilhelm Westphal listed in the bibliography.

There is a peak in the radiometric force at a distinct pressure. The force goes down if you go above or below this pressure. The exact value of the pressure where the force peaks differs according to the size and shape of the radiometer bulb and the vanes, as well as how close the vanes are to the bulb walls. For the common sized radiometers that you can buy today, the pressure is on the order of 0.1 mmHg (13 Pa).

Crookes published data showing this peaking phenomenon for the case of a radiometer with aluminum vanes in the shape of cups 10 mm in diameter. Figure 24 is a plot of this data as a function of pressure in units of pascals. The vertical axis is the rotation speed of the vanes in revolutions per minute which is directly related to the force on the vanes. You can see from the plot that there is a peak at a pressure of about 4 Pa. It drops off sharply for lower pressures and more gradually for higher pressures.

In the work of Wilhelm Westphal (see bibliography), he plots the force divided by the peak force versus the logarithm of the pressure divided by the peak pressure. The

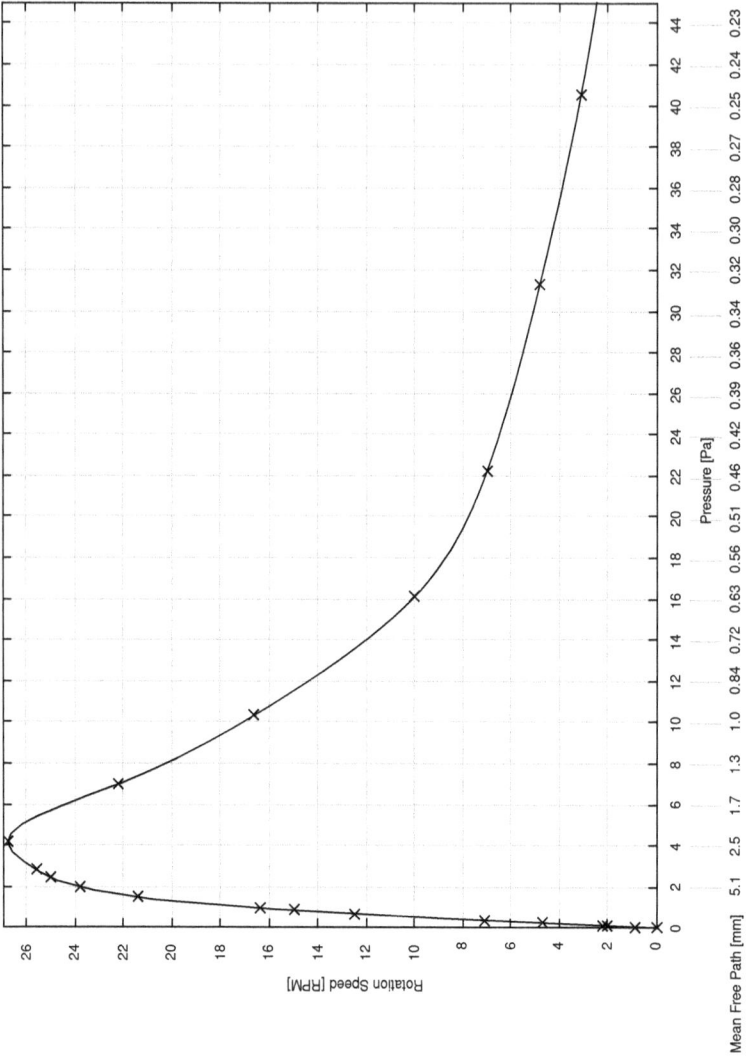

Figure 24: Rotation speed (RPM) vs. pressure (pascals) with cup shaped vanes, from Crookes' published data.

result is a perfectly symmetric bell shaped peak. Let F_0 be the peak force and P_0 be the peak pressure then Westphal calls the function

$$\frac{F}{F_0} = f\left(\log \frac{P}{P_0}\right)$$

the radiometer function but he does not specify the mathematical form of this function. The mathematical form was later proposed by Brüche and W. Littwin (see bibliography), it is discussed in a later section (see: Empirical formula for radiometric force as a function of pressure).

A very recent measurement of the radiometric force versus pressure was done by Ben Krasnow on his Applied Science YouTube channel. You can see a video of this at: https://www.youtube.com/watch?v=r7NEI_C9Yh0 where he shows a plot at around 4:50. He found that the peak in the rotation speed was at around 7 mTorr which is about 1 Pa.

How the Radiometric Force is Generated

So what exactly is going on here? Why this complicated dependence of the radiometric force on the pressure? How exactly is the residual gas causing the vanes to turn? These are not easy questions to answer. To answer them in depth would require a deep foray into the kinetic theory of gases which is beyond the scope of this book.

We will give a qualitative, or at best semi-quantitative

description of how the radiometric forces in a given radiometer are generated. A detailed quantitative description would, among other things, have to take into account the temperature distribution inside the radiometer, which is a function of radiative heat transfer from sources outside the bulb, as well as from the bulb body itself. It is also a function of heat conduction in the vanes as well as in the gas. This is a hard problem to solve and in recent years people have used computer simulation in an attempt to understand the forces.

The Radiometer as a Heat Engine

Trying to describe the behavior of the radiometer is somewhat similar to describing the behavior of a particular heat engine. By understanding the thermodynamic cycle a heat engine goes through, we can get a general qualitative idea for how it will operate. A quantitative description however, requires empirical modeling and sometimes computer simulation.

The radiometer is itself a heat engine. Radiant energy flows into the bulb, heating the vanes, the black side more than the white. The heat is transferred to the residual gas which is the working fluid in this engine. The gas exerts force on the vanes causing them to turn. We could say that this is the work produced by the engine but in most radiometers the work only serves to increase the internal energy of the gas. Crookes did, however, create a radiometer with a magnetized needle attached to one of the arms. As the arm rotated, it made a magnet outside the bulb oscillate, so it is possible to extract mechanical work from the turning of the arms.

Besides the pressure in the bulb, the most important parameter in this heat engine is the temperature difference between the black and white sides. When both sides have the same color, they will have the same temperature, and there will be no movement.[27] It is the temperature difference that drives the dynamics of the residual gas and this creates the force that turns the vanes.

The same is true for radiometers with vanes that are not flat, such as the cup shaped vanes Crookes experimented with. These vanes have no coloring and can be made out of a single material such as aluminum foil. When exposed to light they turn just as well as the flat vaned radiometers. They are also powered by temperature differences between different parts of the vanes. Due to their shape, they are not heated uniformly by the incident light.

Without temperature gradients inside the bulb there could be no net flow of gas and no pressure differences so the vanes would not turn. To maintain the temperature gradients there must be a low temperature reservoir where the gas can give up heat. This is true for any heat engine. In the radiometer, the glass bulb in contact with the outside environment is the low temperature reservoir.

If there was no way for the gas to give up heat, everything

---

[27]Crookes did show that it was possible to get movement in a radiometer where the vanes had the same color on both sides, but this involved creating a temperature gradient in the bulb, which had the same effect as a temperature difference between the sides. We will not consider this type of radiometer.

in the bulb would eventually reach a uniform temperature and the pressure in the bulb would rise. Eventually, the vanes would stop turning. In normal steady state operation, without any external work being produced, the radiant energy flowing into the bulb must equal the heat flowing out.

In some experiments Crookes was able to make the radiometer run in reverse with the black side leading. This can only happen when the black side becomes cooler than the white. The black side, being a good absorber of radiation, is also a good emitter of infrared radiation. This means that if the bulb becomes cold enough and the inflow of radiant energy is low enough, then the black side will lose more energy to the bulb and cool off faster than the white side.

Once the black side becomes cooler than the white, the radiometer will run in reverse. An easy way to make this happen is to put the radiometer in a freezer and partially close the door. The black side of the vanes will quickly cool off and become colder than the white side. The vanes will then turn opposite to the way they normally do in bright light and they will continue to do so until both the white and black sides reach the same temperature. Take the radiometer out of the freezer and it will very quickly start turning in the usual way.

The basic thermodynamics of the radiometer is pretty clear. However, it still does not tell us exactly how the gas exerts force on the vanes. Let's see if we can get at least a qualitative picture of how these radiometric forces are produced.

A Simple Explanation

If pressed to explain how the residual gas generates the radiometric force, a lot of physicists would probably answer as follows. Since the black side is hotter, any molecule recoiling from it would do so with greater momentum than from the white side. The sum of all these molecular collisions will create a greater force on the black side than on the white side, thus making the vanes turn the way they do, with the white side leading. This is a nice simple explanation, and it was actually the accepted explanation, in the very early days of the radiometer, but people soon began to find problems with it.

The see the problem, consider a cylinder filled with gas where one end is kept at a higher temperature than the other. Using the same argument given above, the hotter side would have a greater force on it than the colder side and the entire cylinder would be propelled in the direction of the hot side. It would be propelled without any external force, which violates the basic laws of mechanics.

The fact is that temperature gradients can exist in a container of gas at equilibrium but pressure gradients usually cannot. In the absence of a force such as gravity, a pressure gradient would cause the gas to redistribute itself, so as to equalize the pressure. What does vary in the gas is the number density or the number of molecules per unit volume. The number density is related to the pressure and temperature as follows:

$$n = \frac{N}{V} = \frac{P}{kT}$$

where N is the number of molecules, V is the volume, P is the pressure, T is the temperature in Kelvin, and k is Boltzmann's constant. With the pressure constant, the number density is inversely proportional to the temperature. In the cylinder example, the density will be lowest at the hot end and highest at the cold end. This density variation means that, even though there is more momentum per molecule at the hot end, there are fewer molecules, so the overall force on the hot end, is the same as on the cold end.

A kinetic explanation for this is that the high energy molecules, recoiling from the hot end, are more effective at scattering away approaching molecules. This results in fewer molecules hitting the hot end as opposed to the cold end, so the force balances out.

This same screening effect should also occur on the radiometer vanes. Molecules recoiling from the warmer black side of the vane, should scatter away enough approaching molecules to balance out the force from the molecules recoiling from the white side. But if this were true then there would be no force on the vanes and they would not turn. But they do turn, so what is going on?

The thing that must be taken into account is that the vane is a thin, finite, flat surface suspended in the gas. Near the center of the black side of the vane it is true that there are enough high energy molecules, recoiling from the surface, to effectively scatter away enough of the approaching molecules to just balance the force on the white side. Near the edge, however, there are fewer of these high energy molecules so the molecules approaching the edge are less effectively scattered. The forces on the two sides of the vane therefore cancel out near the center but not near the edge and so the vane moves.

If this explanation is correct it means the force making the radiometer turn should depend on the edge length of the vanes. There is some experimental evidence to support this hypothesis. A paper by Marsh, Condon, and Loeb[28] published in 1925 looked at the relationship between edge length and the force on the vanes. In particular, they looked at the force on vanes such as the one in figure 25.

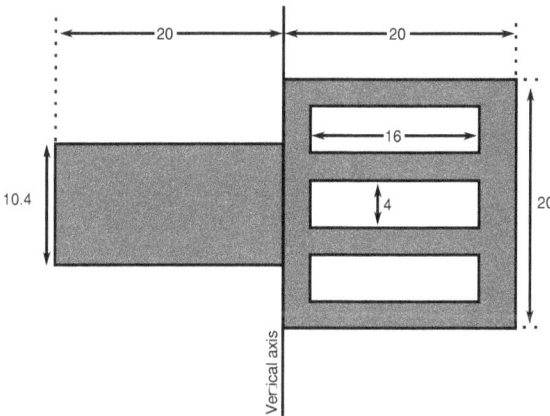

Figure 25: Left and right vane halves have same area, but different total edge lengths.

The two halves of the vane have the same area but the right half has three rectangular holes giving it a much greater edge length. The right half has 2.62 times the

---

[28]The theory of the radiometer, H.E. Marsh, E. Condon, and L.B. Loeb, Journal of the Optical Society of America and Review of Scientific Instruments, Vol 11, No 3, Sept 1925, p257-262.

edge length of the left half. The two halves have equal area moments but the edge moment of the right half is about 3.3 times greater than the left half (see problem section). The vane was made of 0.045 mm thick mica with one side blackened. The vane was attached to a fine quartz fiber so that it could rotate about the line marked "vertical axis" in the figure.

When the entire blackened side of the vane was illuminated, it turned counter clockwise when looking down the axis of rotation. This showed that the right half had a greater force on it than the left. When each half was illuminated separately it still showed that the right side experienced more force than the left. The difference was not, however, as great as one would expect from the ratio of the edge moments. One thing that could account for this is that the holes decrease the temperature difference between the two sides. Another factor is that the reduced edge screening is not as great for the edges along the holes in the vanes. The experiment does provide strong evidence that the forces are a function of the edge lengths. The greater the edge length, the greater the force.

One question that still needs to be answered is how the force depends on the pressure. If the pressure is either too high or too low there is not enough force to turn the vanes. But the important thing to focus on here is the density of the gas and not the pressure. It is the number of molecules per unit volume that determines whether the vanes will turn or not. As we stated above, the number density is related to the pressure and temperature, $n = P/kT$.

The reason the density is so important is that it determines what is called the mean free path of the molecules, which we will write as λ. This is the average distance a molecule travels before colliding with another molecule. The equation for λ is

$$\lambda = \frac{1}{n\sigma\sqrt{2}}$$

where σ is the effective cross-sectional area of a molecule. If d is the diameter of a molecule then $\sigma = \pi d^2$. The diameters are on the order of angstroms (1 angstrom = 0.1 nanometer). For the oxygen and nitrogen molecules that are the primary constituents of air, d is about 3 angstroms.

The significance of λ is that it determines the distance from the edge over which the radiometric force acts. Only for distances from the edge that are less than a few times λ will the number of recoiling molecules be too few to scatter away enough of the approaching molecules to balance the force on the other side. So the edge force driving the radiometer acts on a strip along the edge only a few λ's in width.

As the density increases, λ decreases and the strip gets smaller so the force gets smaller. At some point it gets too small to generate enough force to turn the vanes. As the density decreases λ increases so the edge strip where the force is generated gets larger and eventually covers the whole vane. But there is a trade off here. The lower density means there are fewer molecules recoiling off the vane. So the force area gets larger but the number of molecules gets smaller. If the number of molecules gets too small, they can't generate enough force to turn the vanes.

It is clear that there must be an optimal density, where the number of molecules recoiling from the vane, together with the width of the edge strip, will generate the maximum force and turn the vanes with maximum velocity. This is where the peak in the force versus pressure graph comes from. If you decrease the density below this point, the width of the edge strip will increase but there are fewer molecules so less force. If you increase the density, there will be more molecules but the width of the edge strip gets smaller so you also get less force.

Thermal Creep Force

There is another possible force producing mechanism in the radiometer that was first investigated experimentally by Osborne Reynolds[29]. His experiments were part of his attempt to more completely understand the forces in the radiometer. He wanted to see if he could get gas to flow in the presence of a temperature gradient, such as exists between the sides of a radiometer vane.

He could not do this at the low pressures required to move the vanes in a usual size radiometer. He knew that shrinking down the vanes would allow him to do the measurements at higher pressures but he could not shrink them down far enough for the pressures he wanted to use. He finally realized that he could do the experiment by look-

---

[29]On certain dimensional properties of matter in the gaseous state, Osborne Reynolds, Philosophical Transactions of the Royal Society, Vol 170, 1879, p727-845.

ing at the motion of gas through a porous plug with a temperature gradient across it.

In his experiments he had two gas chambers separated by a porous plate. For the plate he used unglazed white porcelain, meerschaum and stucco. Keeping one of the chambers hot and the other cold he found that gas would flow from the cold to the hot chamber through the plate. Starting out with the same pressure in each chamber, he would end up with a higher pressure in the hot chamber. He called this flow through the plate from hot to cold, thermal transpiration. He also found a simple relationship between the ratio of pressures and temperatures in the two chambers.

$$\frac{P_1}{P_2} = \sqrt{\frac{T_1}{T_2}}$$

This relationship can be derived from kinetic theory. It comes from the fact that the molecular flux in a gas at pressure, P and temperature, T is proportional to $P/\sqrt{T}$. At equilibrium the flux from cold to hot through a pore must equal the flux from hot to cold, so we have $P_1/\sqrt{T_1} = P_2/\sqrt{T_2}$. For this to be valid the mean free path must be large enough so that molecules traveling through the pore interact mostly with the walls and not with each other.

The Danish physicist Martin Knudsen found this same relationship when he studied the flow of rarefied gases through tubes early in the twentieth century. He used tubes with diameters much larger than the pores in the plates Reynolds used but the pressures were so low that the mean free path was much larger than the radius of the tube. He found remarkable agreement with the above equation. In one of his experiments[30] he measured $\frac{P_1}{P_2} = 1.320$ and $\sqrt{\frac{T_1}{T_2}} = 1.329$. The difference could hardly be distinguished from experimental error. Knudsen calls the pressure difference between the hot and cold ends of the tube, the thermal molecular pressure. He describes a simple way to directly observe this pressure.

"The thermal molecular pressure can be demonstrated by means of a bulb made of porous porcelain. The air inside the bulb is heated by passing an electric current through a resistance. When a steady state is reached the inner surface of the porcelain is hotter than the outer surface. This causes a flow of air through the porcelain from the cool side to the hot, i.e. from the outside to the inside. The air pressure inside thus augments and the air can be made to pass through a tube, the end of which dips into water. A continual bubbling is then observed, and if the air be collected in a reversed beaker filled with water, the audience will see that the air must have come through the pores of the porcelain and is not caused merely by an expansion of the air in the bulb."[31]

The theoretical description of thermal transpiration was further developed by Maxwell[32]. He showed that thermal transpiration can take place along a surface on which there

[30] The Kinetic Theory of Gases, Martin Knudsen, 1934.

[31] Knudsen, p.35-36.

[32] On stresses in rarified gases arising from inequalities of temperature, James Clerk Maxwell, Philosophical Transactions of the Royal Society, Dec 31 1879, Vol 170, p231-256.

is a temperature gradient. This phenomenon is called thermal creep flow. As the gas flows along the surface from low to high temperature it exerts a force on the surface that is opposite to the direction of flow. This force on the surface, in the direction from hot to cold, is called the thermal creep force.

Maxwell developed a general set of differential equations describing how a gas interacts with a surface that has a temperature gradient. Unfortunately the equations are difficult to solve and have only been solved for a very limited number of cases.

An example of what the thermal creep flow for flat and cup vane radiometers may look like is shown in figure 26. For the common flat vaned radiometers, the thermal creep flow probably contributes very little to the motion since there are significant thermal gradients only across the edges from one side to the other and these are very thin.

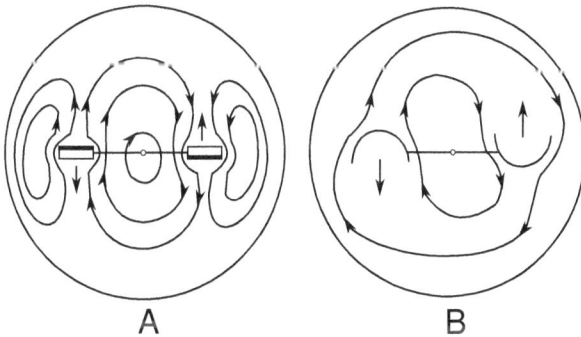

Figure 26: Thermal creep flow for flat and cup vane radiometers (from Kinetic Theory of Gases, Earle H. Kennard, 1938, pg. 344).

For some kinds of radiometers, however, the thermal creep force can be quite significant. A paper published in 2016[33] looked at a radiometer where the only force driving the vanes was the thermal creep force. Instead of vertical vanes with one side black and the other white, the vanes were horizontal with half of each side colored white and the other half black. When light shines on the vane, the black half heats up more than the white. This sets up a temperature gradient and causes a thermal creep flow of gas along the surface from the white to the black half of the vane. The force exerted on the vane is opposite to the flow direction so that the vanes turn with the white halves leading.

A video of these radiometers in action can be found at the URL:
https://calhoun.nps.edu/handle/10945/56287

These radiometers show that the thermal creep force is very real and can be quite significant. The paper compares experimental, theoretical, and simulation results. The theoretical and simulation results did not seem to match the experimental results very well. It indicates that more work on theoretical modeling of this phenomenon may be necessary.

---

[33] A horizontal vane radiometer: experiment, theory, and simulation, David Wolfe, Andres Larraza, Alejandro Garcia, Physics of Fluids, Mar 14 2016, Vol 28, No 3, article 037103.

Empirical formula for radiometric force as a function of pressure

As the pressure is increased from a perfect vacuum, the radiometric force increases sharply from 0 to a peak at a pressure of around 1 pascal. Past the peak, the force decreases with increasing pressure and gradually goes to zero. This suggests that for pressures well below the peak, the force is directly proportional to pressure and for pressures well above the peak, the force is inversely proportional to pressure. Below the peak we have $F \approx P/a$ and above the peak we have $F \approx b/P$.

In 1929 E. Brüche and W. Littwin [34] published a paper where they suggested the following empirical formula for the radiometric force as a function of pressure

$$F(p) = \frac{1}{\frac{a}{p} + \frac{p}{b}}$$

This formula is the harmonic mean of the two approximations given above for the force below and above the maximum. The constants a and b can be fit to the data. The function has a maximum of $F_0 = \frac{b}{2\sqrt{ab}}$ which occurs when the pressure is $p_0 = \sqrt{ab}$ (see problem section). If we let $y = F/F_0$ and $x = p/p_0$ then the function can be written in the following more convenient, unitless form

[34]Experimentelle Beitrage zur Radiometerfrage, E. Bruche, W. Littwin, Zeitschrift für Physik, volume 52, pg 318–333 (1929).

$$y = \frac{2x}{x^2 + 1}$$

Figure 27 shows how well this function fits Crookes' data. There is an overall qualitative fit. Some of the discrepancy could be due to the uncertainty in the position of peak. Some of it could be due to the fact that Crookes was using cups and not flat vanes. For the usual flat vaned radiometers the formula does fit the data quite well[35].

Brüche and W. Littwin also suggested the following generalized version of their formula where the parameter n could possibly be used to improve the fit.

$$F(p) = \frac{1}{\sqrt[n]{\left(\frac{a}{p}\right)^n + \left(\frac{p}{b}\right)^n}}$$

---

[35]See: Experimentelle Beitrage zur Radiometerfrage, E. Bruche, W. Littwin, Zeitschrift für Physik, volume 52, pg 318–333 (1929).

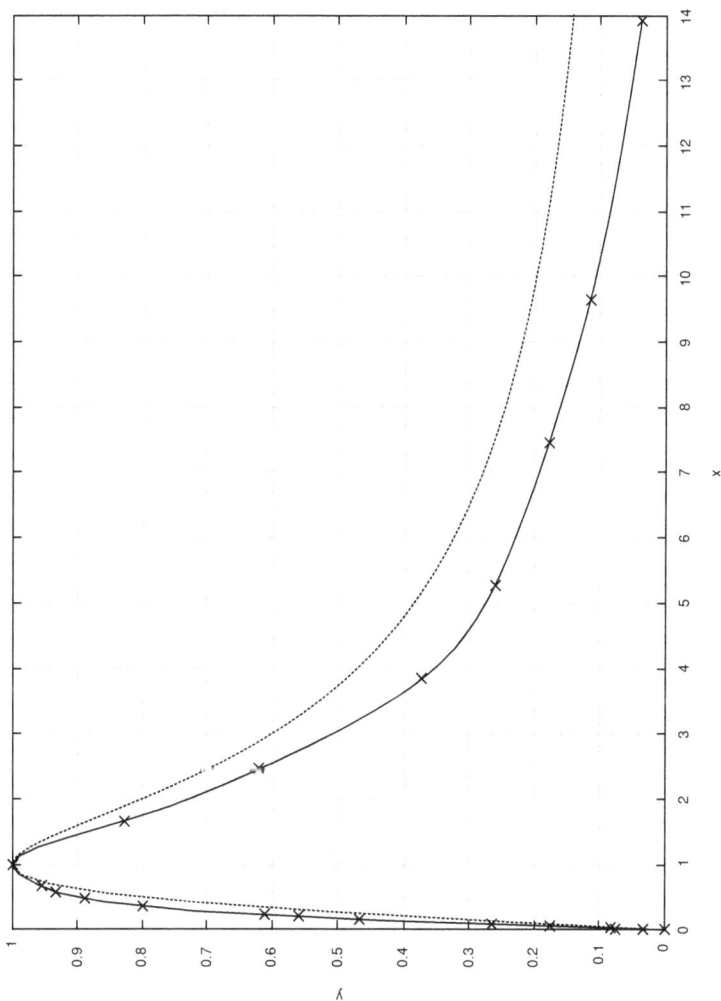

Figure 27: Unitless version of plot in figure 24 with the empirical function of Brüche and Littwin.

Problem 1. The vanes in a radiometer are often square shaped as shown in figure 28.

a) What is the moment of inertia of such a vane about the axis shown in the figure?

b) What is the moment of inertia about a parallel axis a distance $a/\sqrt{2}$ from the one shown in the figure?

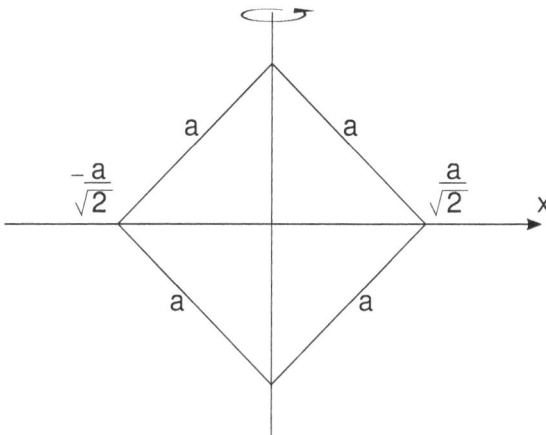

Figure 28: Square shaped radiometer vane of side a rotating about center.

Answer. a) The moment of inertia can be found from the integral

$$I = \int r^2 \, dm$$

where r^2 is the square of the distance of the mass

77

element dm from the rotation axis. As mass elements, take strips parallel to the rotation axis. These strips will have width dx and length

$$l = \begin{cases} 2(\frac{a}{\sqrt{2}} + x) & -\frac{a}{\sqrt{2}} \le x \le 0 \\ 2(\frac{a}{\sqrt{2}} - x) & 0 \le x \le \frac{a}{\sqrt{2}} \end{cases}$$

and mass $dm = \rho t l dx$ where ρ = density of the vane material and t is the thickness. The moment of inertia is then

$$I = \int_{-\frac{a}{\sqrt{2}}}^{\frac{a}{\sqrt{2}}} x^2 \rho t l \, dx$$

Because the vane is symmetric about the axis of rotation, we need only integrate over one side of the vane, then multiply the integral by 2:

$$I = 2\rho t \int_0^{\frac{a}{\sqrt{2}}} 2x^2 (\frac{a}{\sqrt{2}} - x) \, dx$$

$$= 4\rho t \left[\frac{a}{3\sqrt{2}} x^3 - \frac{x^4}{4} \right]_0^{\frac{a}{\sqrt{2}}}$$

$$= \frac{\rho t a^4}{12}$$

$$= \frac{Ma^2}{12}$$

where the last line comes from the fact that the density ρ is mass per volume, and the volume in this case is $a^2 t$.

b) The parallel axis theorem says that the moment of inertia about an axis parallel to an axis that intersects the center of mass is given by $I = I_{cm} +$

Md^2 where I_cm = moment of inertia about the axis through the center of mass, and d = perpendicular distance between the two axes. So for $d = a/\sqrt{2}$ we have:

$$I = \frac{Ma^2}{12} + M\left(\frac{a}{\sqrt{2}}\right)^2$$
$$= \frac{7Ma^2}{12}$$

Problem 2. Radiometer vanes are sometimes circular. Find the moment of inertia of a circular disk of radius r, rotating about an axis offset by a distance a from the edge of the disk as shown in figure 29.

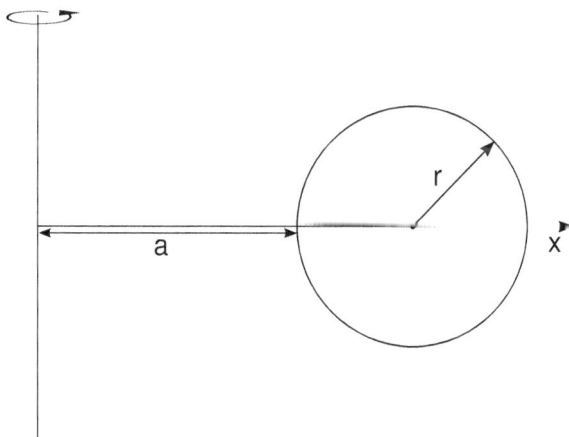

Figure 29: Round radiometer vane of radius r, rotating about an axis offset by a distance a from the edge of the disk.

Answer. We will first get the moment of inertia about
the center of mass, then use the parallel axis theorem to get it at a distance $a + r$ from the center.
Figure 30 shows the disk rotating about its center, divided into vertical strips of length "l" and
infinitesimal width dx. The length of the strips

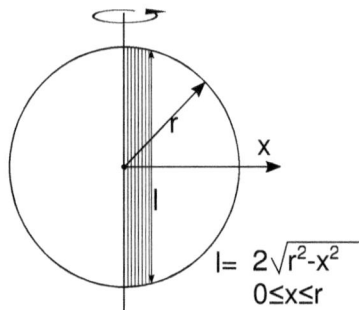

Figure 30: Disk rotating about its center, and decom-
posed of strips of length "l".

shown in figure 30 is $l = 2\sqrt{r^2 - x^2}$ over the range
$0 \leq x \leq r$. The moment of inertia is gotten from
the integral

$$I = \int x^2 \, dm$$

where x^2 is the square of the distance of the mass
element dm from the rotation axis, and $dm = \rho t l dx$
where ρ = density of the vane material, and t is the
thickness. The moment of inertia is then

$$I = 2\rho t \int_0^r x^2 2\sqrt{r^2 - x^2} \, dx$$

Here, because the vane is symmetric about the axis
of rotation, we are only integrating over one side of

the vane, and multiplying the integral by 2.

$$I = 4\rho t \int_0^r x^2 \sqrt{r^2 - x^2} \, dx$$

Now we transform the integral into trigonometric form in order to solve it easier. From the figure:

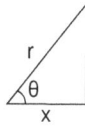

$$x = r \cos \theta$$
$$dx = -r \sin \theta \, d\theta$$

Substituting this into the last integral, and changing the limits according to the new variable θ:

$$\int_0^r x^2 r \sqrt{1 - (x/r)^2} \, dx =$$
$$\int_{\pi/2}^0 r^3 \cos^2 \theta \sqrt{1 - \cos^2 \theta} (-r \sin \theta) \, d\theta =$$
$$r^4 \int_0^{\pi/2} \cos^2 \theta \sin^2 \theta \, d\theta$$

The moment of inertia is then:

$$I = 4r^4 \rho t \int_0^{\pi/2} \cos^2 \theta \sin^2 \theta \, d\theta$$

This integral can be looked up in a table, giving:

$$I = \frac{r^4 \rho t \pi}{4} = \frac{r^2 M}{4}$$

where the last equality is gotten from the fact that the density ρ is mass per volume, and the volume in this case is $\pi r^2 t$. Now we use the parallel axis theorem to get the moment of inertia a distance $r + a$ from the center of the disk:

$$I = I_{cm} + Md^2$$

$$I = M \left(\frac{r^2}{4} + (r + a)^2 \right)$$

Problem 3. Find an expression for the time it takes the radiometer vanes to stop turning if there is only a frictional force in the pivot.

Answer. Let I be the moment of inertia of the vanes, and τ be the torque due to the frictional force. Then the following equation describes the dynamics:

$$I \frac{d\omega}{dt} = -\tau$$

where ω is the angular velocity of the vanes. Rewrite the equation as

$$d\omega = -\frac{\tau}{I} dt$$

Integrating this, you get:

$$\omega = \omega_0 - \frac{\tau t}{I}$$

where ω_0 is the starting angular velocity. The rotation stops when $\omega = 0$, and this occurs when:

$$t = \frac{\omega_0 I}{\tau}$$

Problem 4. Find an expression for the time it takes the radiometer vanes to stop turning if there is a frictional force in the pivot, and a drag force.

Answer. Let I be the moment of inertia of the vanes, τ be the torque due to the frictional force, and a be the linear drag coefficient. Then the following equation describes the dynamics:

$$I\frac{d\omega}{dt} = -\tau - a\omega$$

where ω is the angular velocity of the vanes. Rewrite the equation as

$$-\frac{I}{\tau}\int_{\omega_0}^{\omega}\frac{d\omega}{1 + \frac{a}{\tau}\omega} = t$$

Integrating this, you get:

$$t = \frac{I}{a}\ln\left(\frac{1 + \frac{a}{\tau}\omega_0}{1 + \frac{a}{\tau}\omega}\right)$$

where ω_0 is the starting angular velocity. The rotation stops when $\omega = 0$, and this occurs when:

$$t = \frac{I}{a}\ln\left(1 + \frac{a}{\tau}\omega_0\right)$$

Problem 5. In William Crookes's fifth paper, published in 1878 [36], he varies the pressure in a radiometer while measuring rotation speed, and finds a peak

---

[36] The Bakerian Lecture - On repulsion resulting from radiation - Part V, Philosophical Transactions of the Royal Society, Vol 169, 1878, p243-318.

in rotation speed at a pressure of 41.5 millionths of an atmosphere. What is this pressure in terms of pascals and mmHg?

Answer. From the "Physics Reference" section in the appendix of this book, 1 atm = 101,325 Pa, so

$$41.5 \cdot 10^{-6} \, \text{atm} \cdot \frac{101325 \, \text{Pa}}{1 \, \text{atm}}$$
$$= 4.20 \, \text{Pa}$$

and 1 atm = 760 mmHg, so

$$41.5 \cdot 10^{-6} \, \text{atm} \cdot \frac{760 \, \text{mmHg}}{1 \, \text{atm}}$$
$$= 0.0315 \, \text{mmHg}$$
$$= 31.5 \, \mu\text{mHg}$$

Problem 6. What altitude above the surface of the Earth does the pressure in the previous problem corre-spond to?

Answer. To keep things simple we'll use the barometric formula that assumes a constant temperature as you go up in altitude. That formula is

$$P = P_0 \, e^{-\frac{Mgz}{RT}}$$

Solving this for the altitude, z, we get

$$z = -\frac{RT}{Mg} \ln \frac{P}{P_0}$$

where R is the universal gas constant, $T = 300$ K is the temperature in Kelvin, $M = 0.0289644$ kg/mol is the molar mass of air, $g = 9.8 \ \mathrm{m/s^2}$ is the gravitational acceleration, $P = 4.2 \ \mathrm{Pa}$ and $P_0 = 101325 \ \mathrm{Pa}$ is the pressure at sea level. These numbers give $z = 88600$ meters.

We can also look up the pressure in the standard atmosphere table where $P = 4.2 \ \mathrm{Pa}$ corresponds to $z = 71291$ meters. So taking the average of this and the calculated value we can say that the altitude is on the order of about 80 kilometers. The altitude at which the international space station orbits is approximately 400 kilometers so this is about one fifth of the way there.

Some people have proposed that 80 kilometers be used as the definition of the altitude where space begins. This is known as the von Karman line. Above this altitude it is impossible for aircraft to fly because they would have to travel at faster than orbital velocity to generate enough lift. Satellites cannot orbit below this altitude because the drag force would quickly bring them down.

For more information on the barometric formulas see:
https://en.wikipedia.org/wiki/Barometric_formula

Problem 7. What is the mean free path at the pressure $P = 4.2 \ \mathrm{Pa}$ and temperature $T = 300 \ \mathrm{K}$?

Answer. The formula for the mean free path is

$$\lambda = \frac{1}{n\sigma\sqrt{2}} = \frac{k_B T}{\sqrt{2}P\pi d^2}$$

where n is the number of molecules per volume and σ is the collision cross section of a molecule. If the diameter of a molecule is d then $\sigma = \pi d^2$. For air molecules $d \approx 3 \cdot 10^{-10}$ m so $\sigma \approx 2.8274 \cdot 10^{-19}$ m^2.

The formula for n in terms of temperature and pressure is $n = \frac{P}{k_B T}$ where $k_B = 1.380649 \cdot 10^{-23}$ J·K^{-1} is Boltzmann's constant. So for $P = 4.2$ Pa and $T = 300$ K the number density is $n = 1.0140 \cdot 10^{21}$ molecules/m^3 and the mean free path is $\lambda \approx 0.0025$ m or about a quarter of a centimeter. Compare this to the mean free path in air at sea level and room temperature, which is about 60 nanometers, or $1/42000$ smaller.

Problem 8. How many molecules are there per cubic millimeter at pressure $P = 4.2$ Pa and temperature $T = 300$ K?

Answer. From the previous problem, the number of molecules per cubic meter is $n = 1.0140 \cdot 10^{21}$ molecules/m^3 so the number per cubic millimeter is $n = 1.0140 \cdot 10^{12}$.

Problem 9. If the probability density function for free path lengths in a gas is $p(r) = e^{-r/\lambda}/\lambda$, where λ is the mean free path length, what is the probability that a molecule will have a free path length greater than λ?

Answer. The probability that the free path length is less than or equal to λ is given by the integral:

$$P(r \le \lambda) = \int_0^\lambda \frac{e^{-r/\lambda}}{\lambda} dr$$

$$= 1 - \frac{1}{e}$$

So the probability the free path length is greater than λ is

$$P(r > \lambda) = 1 - P(r \le \lambda)$$

$$= \frac{1}{e}$$

Since $1/e = 0.367879$, this means that about 37% of the time a molecule will have a free path length greater than the mean free path length.

Problem 10. In a 1925 paper by Marsh, Condon, and Loeb [37] they looked at the force on the vane shown in figure 31. The two halves of the vane have the same area but the right half has three rectangular holes giving it a much greater edge length (2.62 times that of the left half). If each side is illuminated separately and the force is generated only along the edges, we would expect there to be more torque on the right side than on the left. Assuming a constant force per unit length along the edges, show that the right side has 3.289 times the edge moment of the left side, while the area moments of the left and right sides are equal.

[37]The theory of the radiometer, H.E. Marsh, E. Condon, and L.B. Loeb, Journal of the Optical Society of America and Review of Scientific Instruments, Vol 11, No 3, Sept 1925, p257-262.

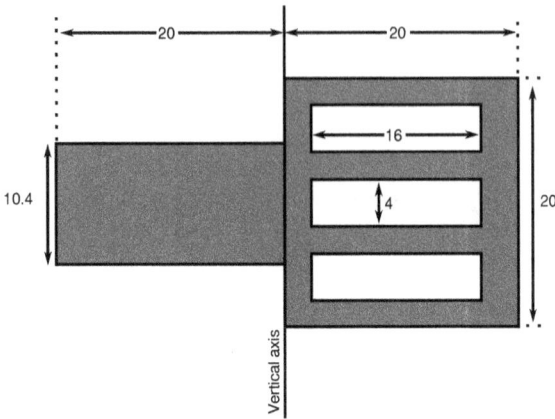

Figure 31

Answer. The edge moment is the edge length times the distance from the axis of rotation. The edge moment of the left side is

$$10.4 \cdot 20 + 2 \int_0^{20} x \, dx = 208 + x^2 \Big|_0^{20}$$

$$= 608$$

The edge moment of the right side is

$$20 \cdot 20 + 2 \int_0^{20} x \, dx + 3 \cdot 4 \cdot 18 + 3 \cdot 4 \cdot 2 + 6 \int_2^{18} x \, dx$$

$$= 400 + 400 + 12 \cdot 20 + 3x^2 \Big|_2^{18}$$

$$= 800 + 240 + 3(18^2 - 4)$$

$$= 2000$$

The ratio of the edge moments is then

$$\frac{\text{right side edge moment}}{\text{left side edge moment}} = \frac{2000}{608} = 3.289$$

The area moment is the area times the distance from the axis of rotation. The area moment of the left side is

$$10.4 \int_0^{20} x\, dx = 5.2x^2 \Big|_0^{20}$$
$$= 2080$$

The area moment of the right side is

$$20 \int_0^2 x\, dx + 8 \int_2^{18} x\, dx + 20 \int_{18}^{20} x\, dx$$
$$= 10x^2 \Big|_0^2 + 4x^2 \Big|_2^{18} + 10x^2 \Big|_{18}^{20}$$
$$= 40 + 4(324 - 4) + 10(400 - 324)$$
$$= 2080$$

So the area moments of the left and right sides are both equal to 2080, giving a ratio of 1.0.

Problem 11. Figure 32 shows a bifilar pendulum suspended from points A and B by wires of length h spaced a distance $2d$ apart. At rest the pendulum lies along the x-axis. If the weight of the pendulum is W show that when it is turned by an angle θ, the restoring torque is given by

$$\tau = Wd \sin \phi \approx \frac{Wd^2}{h} \theta$$

Answer. From figure 32 we have

$$a = 2h \sin(\phi/2)$$
$$a = 2d \sin(\theta/2)$$

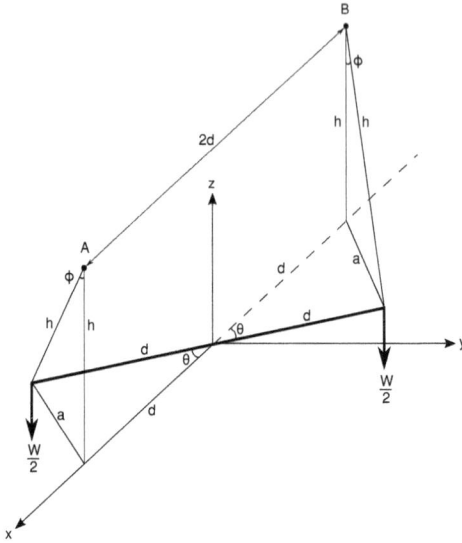

Figure 32

so that
$$h \sin(\phi/2) = d \sin(\theta/2)$$
and in the small angle limit
$$h\phi = d\theta$$
$$\phi = \frac{d}{h}\theta$$

There is a restoring force of $\frac{W}{2} \sin \phi$ on both ends of the pendulum so the total restoring torque is
$$\tau = Wd \sin \phi$$
and in the small angle limit
$$\tau = Wd\phi$$
$$= \frac{Wd^2}{h}\theta$$

Timeline of His Life

- 1832 - Born June 17 in London, England.

- 1848 - Enters Royal College of Chemistry, London.

- 1851 - Becomes Senior Assistant to Professor August Wilhelm Hofmann. Publishes his first scientific paper "On the Selenocyanides".

- 1854 - Leaves the Royal College of Chemistry, and is appointed Superintendent of the Meteorological Department of the Radcliffe Astronomical Observatory at Oxford.

- 1855 - Appointed Teacher of Chemistry at the College of Science, Chester.

- 1856 - Marries Ellen Humphrey on April 10. Becomes editor of the Liverpool Photographic Journal.

- 1857 - Publishes a 60 page "Handbook to the Waxed-Paper Process in Photography". Becomes Secretary of the London Photographic Society and editor of its journal. Elected a Fellow of the Chemical Society.

- 1858 - Retires from the London Photographic Society's editorship and secretaryship. Makes a 2 year

Figure 33: William Crookes at the age of 24, circa 1856, From Fournier d'Albe, E. E. (Edmund Edward). "The Life of Sir William Crookes, O.M., F.R.S." London, England: T. Fisher Unwin (Firm), 1923. Public domain image courtesy of Wikimedia Commons at https://commons.wikimedia.org/wiki/ File:Portrait_of_William_Crookes,_age_24.tiff

agreement to publish articles on photography in the "Photographic News"

- 1859 - Starts a new weekly, "The Chemical News", after acquiring the periodical "Chemical Gazette".

- 1861 - Crookes announces his discovery of thallium on March 5 in a letter, and on March 30 in "The Chemical News".

- 1862 - Exhibits metallic thallium on May 1.

- 1863 - Becomes a Fellow of the Royal Society.

- 1864 - Starts the "Quarterly Journal of Science" with James Samuelson.

- 1865 - Begins investigating the cattle plague problem, and gold amalgamation.

- 1867 - His brother dies on September 22, profoundly affecting his outlook on life.

- 1868 - Suffers from ill health, continues publishing "The Chemical News".

- 1870 - Goes on eclipse expedition to North Africa. Writes book on "Beet-root Sugar". Begins investigating spiritualism.

- 1871 - Writes book: "Select Methods in Chemical Analysis". Continues work on atomic weight of thallium. Becomes director of the Native Guano Company to convert sewage into saleable manure.

- 1873 - Writes papers: "On the action of heat on gravitating masses", and "Attraction and repulsion resulting from radiation".

- 1874 - Publishes book "A Practical Handbook of Dyeing and Calico-Printing".

- 1875 - Writes second paper on the subject of the last, and constructs an instrument he calls a radiometer. An optical instrument maker asks for instructions on making radiometers and offers a royalty. Receives invitations from many English societies to lecture and name his own fee.

- 1878 - Shows his radiometer to the Academie des Sciences in Paris. Publishes first paper on what came to be called cathode rays: "On the illumination of lines of molecular pressure and the trajectory of molecules".

- 1879 - Crookes sells his "Quarterly Journal of Science", which becomes a monthly. He is elected a member of the Athenaeum Club on the proposal of Professor George Stokes. Translates book "Artificial Manures: their chemical selection and scientific application to agriculture" by M. George Ville.

- 1880 - French Academy of Sciences awards him a gold medal and prize of 3,000 francs on March 1. Attends his last board meeting of the Native Guano Company. Moves into large house at 7, Kensington Park Gardens, London: the first house in England lighted by electricity, and was equipped with a laboratory where Crookes and sons worked on developing incandescent lamps.

- 1881 - On February 17 presents to the Royal Society a report on experiments of the viscosity of gases at high vacuum, confirming Maxwell's prediction that the viscosity of a gas is independent of pressure.

- 1884 - On January 16 he loses his father, aged 92.

- 1885 - On June 18 presents to the Royal Society a report on the spectroscopy of phosphorescent light exhibited by chemical substances bombarded by cathode rays.

- 1886 - Elected President of the Chemical Section of the British Association. Mistakenly announces the discovery of 9 new elements by the method of phosphorescent spectra. Completes second edition of book "Select Methods in Chemical Analysis".

- 1888 - Awarded the Davy Medal of the Royal Society. Bought a Hammond typewriter, and at the age of 56, taught himself typewriting, and soon could type faster than he could handwrite.

- 1890 - Elected President of the Institution of Electrical Engineers.

- 1897 - Received the Honour of Knighthood from Queen Victoria. Elected President of the Society of Psychical Research.

- 1898 - Elected President of the British Association for the Advancement of Science.

Figure 34: Sir William Crookes at the age of 57, circa 1889, From Fournier d'Albe, E. E. (Edmund Edward). "The Life of Sir William Crookes, O.M., F.R.S." London, England: T. Fisher Unwin (Firm), 1923. Public domain image courtesy of Wikimedia Commons at https://commons.wikimedia.org/wiki/ File:Portrait_of_William_Crookes,_age_57.tiff

- 1900 - Became Honorary Secretary of the Royal Institution of Great Britain, holding the office for 13 years, arranging the program of lectures every year.

- 1903 - Invented the spinthariscope, which shows flashes of light upon the disintegration of radium atoms.

- 1904 - Awarded the Copley Medal of the Royal Society for work on the effect of cathode rays on various substances.

- 1906 - Received the honour of Correspondent to the Institut de France. Celebrates golden wedding on April 10. Joined Ordnance Research Board as a salaried officer.

- 1908 - Elected Foreign Secretary of the Royal Society. For the Royal Society, wrote obituary of Henri Becquerel, discoverer of radium rays. Lady Crookes contracts pneumonia which leaves her permanently enfeebled.

- 1910 - Awarded the Order of Merit.

- 1911 - Received the Medal of the Society of Chemical Industry. A portrait of Crookes by E. A. Walton was formally presented to the Royal Society.

- 1913 - Elected President of the Royal Society, occupying the office for 3 years. Elected President of the Society of Chemical Industry.

- 1915 - Delivers his last address as President of the Royal Society, ceding the Chair to his successor, Sir J. J. Thompson.

Figure 35: Sir William Crookes at the age of 79, circa 1911, From Fournier d'Albe, E. E. (Edmund Edward). "The Life of Sir William Crookes, O.M., F.R.S." London, England: T. Fisher Unwin (Firm), 1923. Public domain image courtesy of Wikimedia Commons at https://commons.wikimedia.org/wiki/ File:Portrait_of_Sir_William_Crookes,_O.M.,_age_79.tiff

- 1916 - Celebrates diamond wedding on April 10. Lady Crookes dies on May 17.

- 1918 - Completes research paper on the rare metal scandium.

- 1919 - Died April 4 in London at the age of 86.

Sir Oliver Lodge says in the foreword of Edmund Edward Fournier d'Albe's *The Life of Sir William Crookes*, that "Crookes was a great experimental chemist". Through his experimental work he:

- Discovered the element thallium.

- Invented his namesake radiometer which stimulated understanding of the kinetic theory of gasses and radiation exchange.

- Invented the cathode ray tube which he correctly identified as exhibiting "matter in a fourth state", that is, neither solid, liquid nor gas.

- Invented the spinthariscope, which shows flashes of light upon the disintegration of radium atoms.

- Following many years of work on measuring atomic weights, came upon the idea of what would later be called isotopes.

Crookes' theoretical understanding of the radiometer and the direction of his experimental work benefited greatly from his collaboration with Professor George Stokes, another fellow of the Royal Society.

An overview of Sir William Crookes's life follows. It is excerpted from the "Introductory" chapter of *The Life of Sir William Crookes* [38] by Edmund Edward Fournier d'Albe, 1923.

> Crookes is the very type and symbol of English science at its best. His career embodies the emergence of the scientific man as a force in English life. He was not of the governing class. His education could not be epitomised as of "Eton and Christchurch", or "Rugby and Trinity". According to early Victorian standards, he had no rightful part to play in English polity at all. His long life of eighty-seven years saw the advance of science from the humble rank it held in 1832 to the all-embracing position to which it attained in 1919, and his own activities and achievements contributed not a little to the astonishing transformation.
>
> Crookes was no linguist. He had no university education, nor did he hold a professorship. His work was partly journalistic and partly that of a consultant. He stood primarily for the widening and dissemination of chemical knowledge and its application to the manifold problems of human life. But his outlook went far beyond that narrow range. He was a keen fighter, but the enemies he preferred to fight were the enemies of the human race. Thus

---

[38]https://archive.org/details/in.ernet.dli.2015.176056/mode/2up

he enlisted in the ranks against cattle plague and cholera when these pests were raging in England, and rapidly gained the higher command by his industry and keen insight. He studied disinfectants and water supply and sewage disposal, leaving a mark on his generation in the shape of a substantial reduction of the death rate. He threw his whole weight into the development of the photographic art, devised new processes, invented new apparatus, and applied the art to the investigation and recording of scientific phenomena such as meteorological changes and solar eclipses. He even attempted — somewhat prematurely — to photograph a projectile fired from a heavy gun and thus to record its trajectory.

All this was done in the first flush of youthful ardour. But as his experience ripened and his resources increased we find him seeking for hidden treasures at deeper levels. Selenium — an element destined for a career of exceptional interest — had fascinated him from his earliest student days. In examining its spectrum with the newly invented "spectroscope", he found a beautiful green line which nobody had seen before. It belonged to an unknown element, and the story of how Crookes tracked down this substance and triumphantly established its elementary nature in the face of much criticism forms one of the most romantic chapters of his life.

But the strangest interlude in Crookes's career occurred in 1870, when he commenced his four years' investigation of "the phenomena called spiritual". He was then thirty eight, and had been married fourteen years. It was quite a new departure for him. He threw himself into the investigation with his usual energy and resource, and achieved the same prominence as a psychical researcher as he had done as a chemist and physicist. His results, taken at their face value, are the most amazing things ever obtained by a trained man of science, and, if fully accepted by the scientific world, would bring about a revolution in our views of the universe such as has not been witnessed since the days of Copernicus.

It was but an interlude. Crookes found himself struggling with unfamiliar conditions in a turbulent atmosphere, which made it impossible to convince his colleagues of the reality of phenomena which he had established to his own satisfaction in his laboratory. Such a state of things is not unknown even in purely physical investigations, but it almost always signifies some flaw in the reasoning, some neglected source of error. Crookes did not feel justified in sacrificing more years to such a fruitless investigation — fruitless of that appreciation by his peers which is so powerful an incentive towards scientific endeavour. And so Crookes closed that chapter, regretfully perhaps, but fully determined to devote all his strength to ultimate problems of a nature open to accepted scientific methods.

And then came that wonderful chapter of researches in high vacua, leading to "radiant matter", the Radiometer, and the "Crookes tube", which incidentally solved the problem of electric lighting, and is now universally represented by the electric lamp found even in humble homes.

Here we find Crookes at the very height of his career. He was, indeed, the outstanding discoverer of the day and of his generation. Working mostly alone and apart in his great laboratory, he wrested many a secret from Nature and laid bare his hard won treasures before an astonished world. For the next thirty years honours fell thick upon him from all sides. The presidency of the Chemical Society, of the Institution of Electrical Engineers, of the British Association for the Advancement of Science, and finally of the Royal Society fell to him in succession, thus giving him some of the most coveted distinctions open to science in England. The knighthood conferred upon him in 1897 was but the Royal assent to a full measure of recognition already earned and received. With a mind untrammelled by dogmatic preconception and a position independent of academic punctilios, he was free to seek the truth wherever it was likely to be found and to promulgate it without fear or favour. As the Victorian era approached its end, and science gradually advanced to a position of greater influence and importance, the position and authority of a man like Crookes assumed an ever increasing prominence. When finally the European War fell upon the world, the silent revolution completed itself dramatically. The world stood face to face with stark reality. Rhetorical subtleties lost their charm, and men and systems were valued according to the extent to which they harmonised with fundamental truths. It was inevitable that the scientific ideal and outlook should then prevail. For the true man of science worships but one god — truth. He despises the ecclesiastic for teaching half-truths for the sake of moral influence, the politician for dressing up truth in a partisan guise, and the business man for subordinating truth to personal gain. Science represents a new aristocracy based upon a new power, but an ascendancy ouverte aux talents and attainable by anyone who will go through the necessary mental and physical labour. Like all aristocracies, it is "tempered by revolution", and one is not surprised to hear the "tyranny of experts" denounced as the worst tyranny of all. But any such revolution could only replace one expert by another expert of greater reputation for knowledge and honesty, and therefore endowed to a greater extent with the ideal attributes of the scientific mind. We thus get a close approximation to that "benevolent despotism" which has been advocated as the only cure for the evils of democratic government.

Not that Crookes even remotely resembled a despot. In his later years, he gave one the impression of a shrewd and kindly personality, venerable with his white hair and beard. His married life was serene and fruitful, and culminated in a diamond wedding. Up to the last years of his long life he worked away at scientific problems in the hope of a solution which would benefit mankind, and when he attained a solution he gave it freely to the world. While his discoveries created, or assisted in creating, vast industries, he was content to do the work of the pioneer, who, as a rule, is notoriously ill rewarded. And therein he represented but another aspect of the scientific discoverer. While statesmen are at pains to control the forces which together make up the life of the nation, the silent investigator created new forces of incalculable import. While the agitator talks of revolution by violence or by stoppage of work, the chemist or physicist in his "peaceful cell" creates or destroys the work of millions of men, and fundamentally alters the status of both labour and capital. The discoverer is the real arbiter of the destinies of the world. The powers that ruled the Middle Ages knew better than to let him loose. And now it is too late, and we must accept the discoverer's revolutionary activity as we accept the earthquake and the tornado, and statesmen and politicians and Labour leaders must needs dance to the tune of the discoverer's pipe. The alternative is to stop the activities of the discoverer. This, indeed, has often been attempted, by bribery and intimidation and what not else. But it is futile, for the new aristocracy is in itself more democratic than any other human institution. There is no central authority — all scientific authority is provisional. It only exists until displaced by wider knowledge and deeper research. Unlike ecclesiastical authority, it is. adaptable and amenable to new truth. It is intensely alive by that very fact, and is indestructible so long as it retains that adaptability. Kings and armies and financial combinations must bow before it. The new despotism is calmly accepted almost everywhere. It therein somewhat resembles the despotism of fashion which, presumably, is also a "tyranny of experts".

Sir William Crookes would probably have objected to being classed as a revolutionary. Standing at the head of the staircase at Burlington House and extending a Presidential welcome to the Fellows and friends of the Royal Society, he seemed to embody the dignity of an ancient, well tried and conservative institution, the conservatism of which had been painfully brought home to him during the spiritualistic interlude. It is fortunate for mankind that the youthful impetuosity of its budding "despots" is tempered by the mature wisdom of the elder men. The body of scientific workers has, so to speak, automatically secreted a cortex as the sap does in a tree, so as to add weight and permanence to the general structure. Crookes in his career passed through all the stages from sap to cortex, but when the Great War overtook him at the age of eighty two he showed a return to his earlier stages, and served his country right through, retiring from the field only when his country had achieved victory and the land that bore him was ready to clasp him to her breast.

That was the man. And how he arose and lived and fought and won his laurels I shall endeavour to show forth in the chapters here following.

If you would like to read more, the entirety of the book can be found here:
The Life of Sir William Crookes by Edmund Edward Fournier d'Albe, 1923
https://archive.org/details/in.ernet.dli.2015.176056/mode/2up

A newer biography can be found here:
William Crookes (1832-1919) and the Commercialization of Science, William H. Brock, 2008
http://www.worldcat.org/oclc/878705113

One atmosphere is the atmospheric pressure at sea level. In terms of commonly used pressure units, it has the following values:

1	atm	atmosphere
101,325	Pa	pascals
1013.25	mbar	millibar
760	Torr	torr
760	mmHg	millimeters of mercury
29.9212	inHg	inches of mercury
14.696	psi	pounds per square inch

The pascal (Pa) is the SI unit of pressure:

$$1 \text{ Pa} = 1 \ \frac{\text{N}}{\text{m}^2} = 1 \ \frac{\text{kg}}{\text{m} \cdot \text{s}^2} = 1 \ \frac{\text{J}}{\text{m}^3}$$

$$1 \text{ Pa} - \begin{array}{l} 9.87 \ \mu\text{atm} \\ 0.01 \ \text{mbar} \\ 7.50 \ \text{mTorr} \\ 7.50 \ \mu\text{mHg} \ (\mu\text{mHg is also known as microns}) \end{array}$$

Avogadro constant	$N_A = 6.02214 \cdot 10^{23}$	mol^{-1}
Boltzmann constant	$k_B = 1.380649 \cdot 10^{-23}$	J/K
Molar Gas constant	$R = 8.314463$	$\text{J/(K} \cdot \text{mol)}$

Ideal Gas Law

$$PV = Nk_BT$$
$$= \tilde{n}RT$$

where N is the number of molecules and $\tilde{n} = N/N_A$ is the number of moles. The molar gas constant is $R = N_A k_B$.

Density of Molecules $(\mathrm{molecules/m^3})$

$$n = \frac{P}{k_B T} \tag{1}$$

Molecular Flux $(\mathrm{molecules/(m^2 \cdot s)})$

$$\Phi = \frac{1}{4} n \bar{v} \tag{2}$$

where $\bar{v}$ is the average molecular speed.

Gas pressure $(\mathrm{N/m^2})$

$$P = \frac{1}{3} n m \bar{v^2} \tag{3}$$

where m is the molecular mass and $\bar{v^2}$ is the mean square speed.

Mean Free Path

$$\lambda = \frac{1}{n\sigma\sqrt{2}} = \frac{k_B T}{\sqrt{2} P \pi d^2} \tag{4}$$

where d = molecular diameter $\approx 3\mathring{A}$ for O_2 and N_2

For air molecules the mean free path at room temperature $(68°F = 20°C = 293.15K)$ is
$\lambda = 1.43\mathrm{cm}$ at 1 Pa pressure
$\lambda = 0.141\mu m$ at 1 atm pressure

1. 1874 XV. On attraction and repulsion resulting from radiation
   William Crookes
   Philosophical Transactions of the Royal Society, Vol 164, 1874, p501-527

2. 1875 XVIII. On repulsion resulting from radiation - Part II
   William Crookes
   Philosophical Transactions of the Royal Society, Vol 165, 1875, p519-547

3. 1876 XIII. On repulsion resulting from radiation - Parts III & IV
   William Crookes
   Philosophical Transactions of the Royal Society, Vol 166, 1876, p325-376

4. 1878 XV. The Bakerian Lecture - On repulsion resulting from radiation - Part V
   William Crookes
   Philosophical Transactions of the Royal Society, Vol 169, 1878, p243-318

5. 1879 XV. On repulsion resulting from radiation - Part VI
   William Crookes

Philosophical Transactions of the Royal Society, Vol 170, 1879, p87-134

6. On a new form of the 'sprengel' air-pump and vacuum-tap
Charles H. Gimingham
Procedings of the Royal Society of London, Vol 25, Dec 7 1876, p396–402

7. On the forces caused by evaporation from, and condensation at, a surface
Osborne Reynolds
Proceedings of the Royal Society, Vol 22, Jun 18 1874, p401-407

8. On the forces caused by the communication of heat between a surface and a gas; and on a new photometer
Osborne Reynolds
Philosophical Transactions of the Royal Society, Vol 166, 1876, p725-735

9. On certain dimensional properties of matter in the gaseous state
Osborne Reynolds
Philosophical Transactions of the Royal Society, Vol 170, 1879, p727-845

10. On the nature of the force producing the motion of a body exposed to rays of heat and light
Arthur Schuster
Philosophical Transactions of the Royal Society, Vol 166, Mar 23 1876, p715-724

11. On stresses in rarified gases arising from inequalities of temperature
James Clerk Maxwell
Philosophical Transactions of the Royal Society, Dec 31 1879, Vol 170, p231-256

12. The radiometer and its lessons
G. Johnstone Stoney
Nature, Jan 3 1878, Vol 17, No 427, p181-182

13. Maxwell, Osborne Reynolds, and the radiometer
S. G. Brush and C. W. F. Everitt
Historical Studies in the Physical Sciences, Vol 1, 1969, p105-125

14. William Crookes and the radiometer
Arthur E. Woodruff
Isis, Vol 57, No 2, 1966, p188-198

15. The radiometer and how it does not work
Arthur E. Woodruff
The Physics Teacher, Oct 1968, Vol 6, No 7, p358-

363

16. William Crookes and the fourth state of matter
    Robert K. DeKosky
    Isis, Vol 67, No 1, Mar 1976, p36-60

17. Crookes's Radiometers: a train of thought manifest
    Jane Wess
    Notes and Records of the Royal Society, Dec 20
    2010, Vol 64, No 4, p457-470

18. Crookes' radiometer and otheoscope
    Norman Heckenberg
    Bulletin of the Scientific Instrument Society, Sep
    1996, No 50, p40-42

19. The theory of the radiometer
    H.E. Marsh, E. Condon, and L.B. Loeb
    Journal of the Optical Society of America and Re-
    view of Scientific Instruments, Vol 11, No 3, Sept
    1925, p257-262

20. Further experiments on the theory of the vane ra-
    diometer
    H. E. Marsh
    Journal of the Optical Society of America and Re-
    view of Scientific Instruments, Vol 12, No 2, Feb
    1926, p135-148

21. Messungen am Radiometer
Wilhelm H. Westphal
Zeitschrift für Physik, Vol 1, No 1, Feb 1920, p92-100

22. Messungen am Radiometer II
Wilhelm H. Westphal
Zeitschrift für Physik, Vol 1, No 5, Oct 1920, p431-438

23. Messungen am radiometer III. Uber ein Quarzfaden Radiometer
Wilhelm H. Westphal
Zeitschrift für Physik, Vol 4, No 2, Jun 1921, p221-225

24. Zur Theorie der Radiometerkräfte
A. Einstein
Zeitschrift für Physic 27 (1924): 1-6

25. Experimentelle Beitrage zur Radiometerfrage
E. Bruche, W. Littwin
Zeitschrift für Physik volume 52, pages 318–333 (1929)

26. Experimentelle Beitrage zur Radiometerfrage
E. Bruche, W. Littwin
Zeitschrift für Physik volume 67, pages 333–361

(1931)

27. Radiometerdruck und Akkommodationskoeffizient
Martin Knudsen
Annalen der Physik, Vol 398, No 2, 1930, p129-185

28. Radiometric phenomena: from the 19th to the 21st
century
Andrew Ketsdever, Natalia Gimelshein, Sergey Gimelshein,
Nataniel Selden
Vacuum, Vol 86, No 11, May 31 2012, p1644-1662

29. Area and edge effects in radiometric forces
N. Selden, C. Ngalande, S. Gimelshein, E. P. Muntz,
A. Alexeenko, A. Ketsdever
Physical Review E, Vol 79, article 041201, 2009, 6
pages

30. A horizontal vane radiometer: experiment, theory,
and simulation
David Wolfe, Andres Larraza, Alejandro Garcia
Physics of Fluids, Mar 14 2016, Vol 28, No 3, article 037103, 31 pages

31. Concerning the action of the Crookes radiometer
Gordon F. Hull
Am. J. Phys., Vol 16, No 3, Mar 1948, p185-186

32. Running Crooke's radiometer backwards
Frank S. Crawford
Am. J. Phys., Vol 53, No 11, Nov 1985, p1105

33. Dynamic characterization of a windmill radiometer
A Arenas, L Victoria, F J Abellan and J A Ibanez
Eur. J. Phys., Vol 17, No 6, 1996, p331-336

34. Angular velocity control for a windmill radiometer
A Arenas, L Victoria, F J Abellan and J A Ibanez
IEEE Transactions on Education, Vol. 42, No. 2, May 1999

35. Radiometer mystery
Wojciech Dindorf
The Physics Teacher, Vol 40, Nov 2002, p504

36. A new way to demonstrate the radiometer as a heat engine
V. I. Hladkouski and A. I. Pinchuk
The Physics Teacher, Vol 53, 2015, p109

37. Kinetic Theory of Gases: with an introduction to statistical mechanics,
Earle H. Kennard
1st ed, 4th impression, 1938

38. The Kinetic Theory of Gases,
Leonard B. Loeb

2nd ed, 5th impression, 1934

39. William Crookes (1832-1919) and the commercialization of science,
William H. Brock
Routledge, 2008, 586 pages, ISBN-13: 978-0754663225

40. The Life of Sir William Crookes,
Edmund Edward Fournier d'Albe, with foreword by
Sir Oliver Lodge
T. Fisher Unwin Ltd, 1923, 413 pages

About the Authors

Stefan Hollos and **J. Richard Hollos** are brothers and business partners at Exstrom Laboratories LLC (www.exstrom.com) in Longmont, Colorado. They are physicists and electrical engineers by training, and enjoy anything related to math, physics, engineering and computing. In addition, they enjoy creating music and visual art, and being in the great outdoors. They are the authors of the following books:

- **Passive Butterworth Filter Cookbook**

- **Nell: An SVG Drawing Language**

- **Coin Tossing: The Hydrogen Atom of Probability**

- **Creating Melodies**

- **Hexagonal Tilings and Patterns**

- **Combinatorics II Problems and Solutions: Counting Patterns**

- **Information Theory: A Concise Introduction**

- **Recursive Digital Filters: A Concise Guide**

- **Art of Pi**

- **Creating Noise**

- **Art of the Golden Ratio**

- Creating Rhythms

- Pattern Generation for Computational Art

- Finite Automata and Regular Expressions: Problems and Solutions

- Probability Problems and Solutions

- Combinatorics Problems and Solutions

- The Coin Toss: Probabilities and Patterns

- Pairs Trading: A Bayesian Example

- Simple Trading Strategies That Work

- Bet Smart: The Kelly System for Gambling and Investing

- Signals from the Subatomic World: How to Build a Proton Precession Magnetometer

More information on all these books can be found at the website of Abrazol Publishing
www.abrazol.com
where you can also sign up for the Abrazol newsletter.

Acknowledgments

In ordinary life we hardly realize that we receive a great deal more than we give, and that it is only with gratitude that life becomes rich. It is very easy to overestimate the importance of our own achievements in comparison with what we owe to others.

   Dietrich Bonhoeffer, letter to parents from prison, Sept. 13, 1943

We'd like to thank our parents, Istvan and Anna Hollos, for helping us in many ways.

We thank the makers and maintainers of all the software we've used in the production of this book: the Emacs text editor, the Latex typsetting system, Gimp, Inkscape, Gnuplot, Evince, Mupdf, Maxima, gcc, Bash shell, and the Linux operating system.

Thank You

Thank you for buying this book.

If you'd like to receive news about this book and others published by Abrazol Publishing, just go to

http://www.abrazol.com/

and sign up for our newsletter.

www.ingramcontent.com/pod-product-compliance
Lightning Source LLC
Chambersburg PA
CBHW071155200326
41519CB00018B/5233